Nematodes
for
Biological Control
of
Insects

Author

George O. Poinar, Jr.
Division of Entomology and Parasitology
University of California
Berkeley, California

CRC PRESS, INC.
Boca Raton, Florida 33431

Library of Congress Cataloging in Publication Data

Poinar, Jr., George O.
    Nematodes for biological control of insects.

    Bibliography: p.
    Includes index.
    1. Insect control — Biological control. 2. Nematoda.
3. Parasites — Insects. I. Title.
SB933.3.P64   632'.7   78-10541
ISBN 0-8493-5333-5

© 1979 by CRC Press, Inc.

International Standard Book Number 0-8493-5333-5

Library of Congress Card Number 78-10541
Printed in the United States

THE AUTHOR

George O. Poinar, Jr., Ph.D., is lecturer and insect pathologist at the University of California, Berkeley.

Dr. Poinar obtained his B.S. degree, his M.S. in Zoology. and his Ph.D. in Entomology from Cornell University, Ithaca, New York in 1958, 1960, and 1962, respectively.

Dr. Poinar is a member of the Entomological Society of America, American Association of Biological Scientists, American Society of Nematologists, European Society of Nematologists, and the Society for Invertebrate Pathology.

Among other awards and honors, he is a member of Phi Kappa Phi and Sigma Xi and has received National Academy of Science Exchange Fellowships for the Soviet Union and for Romania.

Dr. Poinar has served as a Nematode Specialist for the Food and Agriculture Organization Rhinoceros Beetle Project, Research Fellow at the National Academy of Sciences, a guest lecturer at the University of Amsterdam, a consultant for the Onchocerciasis Control Program in West Africa, and has published over 100 research papers in nematology and insect pathology.

ACKNOWLEDGMENTS

The author expresses his sincere gratitude to Gerard Thomas for his assistance in the preparation of this work, Diana Morris for typing the manuscript, and Roberta Hess for preparing the plates.

Grateful appreciation is also extended to the following persons who supplied photographs or unpublished information for the present work: J. J. Petersen, R. Levy, E. Hansen, J. Hansen, H. Sandner, M. Laird, M. Remillet, M. Golden, J. Lindegren, C. Doncaster, W. Gaugler, G. De Zoeten, R. Levy, V. R. Ferris, J. M. Ferris, and R. Hess.

TABLE OF CONTENTS

Chapter 1

INTRODUCTION

Nematodes are considered one of the most abundant groups of living animals, and although morphologically they are very simple, they have exploited a wide range of diverse habitats.

Most of us are educated to consider nematodes as harmful organisms, attacking plants and parasitizing vertebrates, including man. It is true that most studies on nematodes are done on species that cause some damage to man and his products, but there are other types of nematodes as well. Basically, nematodes can be arranged into several groups depending on their nourishment. Thus there are the free-living microphagous forms, the predatory forms, the plant parasitic forms, the vertebrate parasitic forms, and the parasites of invertebrates. It is the latter category that we are interested in here. Nematodes can parasitize spiders, leeches, annelids, crustaceans, molluscs, and other groups of invertebrates, including insects. When this parasitism results in death of the host, these forms are potential biological control agents. When the hosts happen to be insect pests of man, then the possibility of control becomes still more intriguing.

With the problems that now exist around the use of insecticides, there is interest in all other possible ways of controlling insects. The purpose of this book is to examine in depth those nematodes which are good candidates for the biological control of insects, either alone or in conjunction with other pest management systems.

To be included in this work, a nematode must meet three criteria. First, it must attack insects that are considered pests of man. Second, it must either kill, sterilize, or seriously hamper the development of the insect. Third, the nematode must have been carried through at least one complete generation either in vivo or in vitro in the laboratory to show that it is capable of some degree of mass production.

Each nematode will be discussed in the present work according to the following outline, which may vary depending on the amount of information available: (1) description, (2) bionomics (life cycle and ecology), (3) host range, (4) culture, and (5) application. Unless otherwise noted, all quantitative values of the nematode species listed in the tables are in microns.

As our knowledge progresses and new entomogenous nematodes are discovered, other candidates will join those already cited here, and the list will expand. However, it is hoped that the following work will stimulate interest in the use of nematodes in our continuous effort to combat insect pests.

Chapter 2

CLASSIFICATION OF NEMATODA

There is no complete agreement today as to how nematodes should be classified. Even the general term, Nematoda (or Nematodea), has been challenged with Nemata as a collective name for all nematodes. For some time, many preferred to group the nematodes as a class within the phylum Aschelminthes (ascos-cavity, helminth-worm). (Earlier this was called Nemathelminthes.) However, recent workers have tended to place the nematodes in a separate phylum, the Nematoda.

The following presents a general classification of the nematodes based on a synthesis compiled from many authors, with those groups discussed in the present work cited in italics.

Some general trends can be found in the present classification which includes nematodes of diverse habitats and characters. For instance, the majority of the Adenophorea consists of free-living marine or fresh-water nematodes. This class only contains three plant-parasitic genera in the suborder Dorylaimina and animal parasites in the order Enoplida.

On the other hand, representatives of the class Secernentia show considerably more diversity in habitat and behavior. Comprising the majority of the free-living soil nematodes, the Rhabditida also include insect parasites, mollusc parasites, annelid parasites, and vertebrate parasites. Members of the order Tylenchida are mainly plant parasites, but some parasitize insects and one specialized group parasitizes leeches. Adults of the orders Strongylida, Ascaridida, and Spirurida parasitize vertebrates, whereas representatives of the Oxyuroidea are found in vertebrates and invertebrates.

The purpose of the following table is to present the great diversity of the nematodes and show how the groups discussed in this work fit into the general classification scheme. The entomogenous nematodes discussed in this work are grouped in order of their appearance in the following scheme.

Higher Classification of the Phylum Nematoda

Class Adenophorea
  Subclass Chromadoria (mostly marine forms)
    Order Araeolaimida
      Suborder Araeolaimina
        Superfamily Araeolaimoidea
          Family Araeolaimidae
          Family Diploplectidae
          Family Haliplectidae
        Superfamily Axonolaimoidea
          Family Axonolaimidae
        Superfamily Leptolaimoidea
          Family Leptolaimidae
          Family Bastianiidae
        Superfamily Camacolaimoidea
          Family Camacolaimidae
        Superfamily Plectoidea
          Family Plectidae
          Family Teratocephalidae
      Suborder Tripyloidina
        Superfamily Tripyloidoidea
          Family Tripyloididae
    Order Monhysterida
      Superfamily Linhomoeoidea
        Family Linhomoeidae

      Superfamily Monhysteroidea
        Family Sphaerolaimidae
        Family Monhysteridae
      Superfamily Siphonolaimoidea
        Family Siphonolaimidae
    Order Desmodorida
      Suborder Desmodorina
        Superfamily Spirinoidea
          Family Spirinidae
          Family Microlaimidae
        Superfamily Desmodoroidea
          Family Desmordoridae
        Superfamily Ceramonematoidea
          Family Dasynemellidae
          Family Ceramonematidae
        Superfamily Monoposthioidea
          Family Monoposthiidae
          Family Richtersiidae
      Suborder Draconematina
        Superfamily Draconematoidea
          Family Epsilonematidae
          Family Draconematidae
    Order Chromadorida
      Suborder Chromadorina

Higher Classification of the Phylum Nematoda (continued)

Superfamily Chromadoroidea
   Family Comesomatidae
   Family Chromadoridae
Suborder Cyatholaimina
   Superfamily Cyatholaimoidea
   Family Cyatholaimidae
   Superfamily Choanolaimoidea
   Family Choanolaimidae
   Family Selachinematidae

Order Desmoscolecida
Suborder Desmoscolecina
   Superfamily Desmoscolecoidea
   Family Meyliidae
   Family Desmoscolecidae
   Superfamily Greeffiellidea
   Family Greeffiellidea
Subclass Enoplia (soil, aquatic, plant-parasitic, and animal parasitic forms)

Order Enoplida
Suborder Enoplina
   Superfamily Tripyloidea
   Family Tripylidae
   Family Ironidae
   Superfamily Enoploidea
   Family Leptosomatidae
   Family Oxystominidae
   Family Lauratonematidae
   Family Phanodermatidae
   Family Thoracostomopsidae
   Family Enoplidae
   Superfamily Mermithoidea
   Family Mermithidae
   Family Tetradonematidae
   Family Echinomermellidae
   Family Marimermithidae
   Superfamily Trichuroidea
   Family Trichuridae
   Family Trichinellidae
   Family Cystoopsidae
   Superfamily Dioctophymatoidea
   Family Dioctophymatidae
   Family Soboliphymidae
   Superfamily Muspiceoidea
   Family Muspiceidae
   Family Robertdollfusidae
   Family Phylotainophoridae
Suborder Oncholaimina
   Superfamily Oncholaimoidea
   Family Oncholaimidae
   Family Eurystominidae
Order Dorylaimida
Suborder Dorylaimina
   Superfamily Dorylaimoidea
   Family Dorylaimidae
   Family Tylencholaimidae
   Family Miranematidae
   Family Aporcelaimidae
   Family Nygolaimidae
   Family Nygolaimellidae
   Family Longidoridae

Superfamily Actinolaimoidea
   Family Actinolaimidae
   Family Neoactinolaimidae
   Family Paractinolaimidae
   Family Carcharolaimidae
   Family Trachypleurosidae
   Family Mylodiscidae
Superfamily Leptonchoidea
   Family Leptonchidae
   Family Belonenchidae
   Family Dorylaimoididea
   Family Tylencholaimellidae
   Family Aulolaimoididae
   Family Campydoridae
   Family Encholaimidae
Superfamily Belondiroidea
   Family Belondiridae
   Family Dorylaimellidae
   Family Axonchiidae
   Family Mydonomidae
   Family Roqueidae
   Family Oxydiridae
   Family Swangeriidae
   Family Falcihastidae
   Family Nygellidae
Superfamily Diphtherophoroidea
   Family Diphtherophoridae
   Family Trichodoridae
Superfamily Monochoidea
   Family Mononchidae
   Family Bathyodontidae
Class Secernentia
Order Rhabditida (soil, aquatic, and animal parasitic forms)
Suborder Rhabditina
   Superfamily Diplogasteroidea
   Family Diplogasteridae
   Family Cylindrocorporidae
   Superfamily Rhabditoidea
   Family Rhabditidae
   Family Bunonematidae
   Family Brevibuccidae
   Family Panagrolaimidae
   Family Myolaimidae
   Family Chambersiellidae
   Family Cephalobidae
   Family Rhabdiasidae
   Family Angiostomatidae
   Family Agfidae
   Family Daubayliidae
   Family Alloionematidae
   Family Steinernematidae
   Family Heterorhabditidae
   Family Strongyloididae
   Family Carabonematidae
   Family Syrphonematidae
Superfamily Drilonematoidea
   Family Drilonematidae
   Family Ungellidae
   Family Scolecophilidae
   Family Creagrocercidae

Higher Classification of the Phylum Nematoda (continued)

Family Mesidionematidae
Family Homungellidae
Family Alaninematidae
Family Pharyngonematidae
Order Tylenchida (mostly plant-parasites,
some invertebrate parasites)
Suborder Tylenchina
Superfamily Tylenchoidea
Family Dolichodoridae
Family Tylenchidae
Family Tylenchorhynchidae
Family Pratylenchidae
Family Hoplolaimidae
Family Belonolaimidae
Superfamily Heteroderoidea
Family Heteroderidae
Family Nacobbidae
Superfamily Criconematoidea
Family Criconematidae
Family Paratylenchidae
Family Tylenchulidae
Superfamily Atylenchoidea
Family Atylenchidae
Superfamily Neotylenchoidea
Family Ecphyadophoridae
Family Neotylenchidae
Family Paurodontidae
Family Nothotylenchidae
Family Allantonematidae
Family Myenchidae
Superfamily Sphaerulariodea
Family Sphaerulariidae
Suborder Aphelenchina
Superfamily Aphelchoidea
Family Anomyctidae
Family Paraphelenchidae
Family Aphelenchidae
Family Aphelenchoididae
Family Entaphelenchidae
Order Strongylida (animal parasites)
Superfamily Diaphanocephaloidea
Family Diaphanocephalidae
Superfamily Strongyloidea
Family Syngamidae
Family Strongylidae
Superfamily Ancylostomatoidea
Family Ancylostomatidae
Superfamily Trichostrongyloidea
Family Amidostomatidae
Family Strongylacanthidae
Family Trichostrongylidae
Family Heligmosomatidae
Family Ollulanidae
Family Dictyocaulidae
Superfamily Metastrongyloidea
Family Metastrongylidae
Order Ascaridida (animal parasites)
Superfamily Cosmocercoidea
Family Cosmocercidae
Family Atractidae
Family Kathlaniidae

Superfamily Seuratoidea
Family Seuratidae
Superfamily Ascaridoidea
Family Acanthoceilidae
Family Anisakidae
Family Ascarididae
Family Crossophoridae
Family Heterocheilidae
Superfamily Heterakoidea
Family Heterakidae
Family Ascaridiidae
Superfamily Subuluroidea
Family Subuluridae
Family Maupasinidae
Order Oxyurida (animal parasites)
Superfamily Oxyuroidea
Family Oxyuridae
Superfamily Coronostomatoidea
Family Cornostomatidae
Family Robertiidae
Superfamily Hystrignathoidea
Family Chitwoodiellidae
Family Aoruridae
Family Pulchrocephalidae
Family Hystrignathidae
Superfamily Thelastomatoidea
Family Thelastomatidae
Superfamily Rhigonematoidea
Family Brumptaemillidae
Family Oniscicolinidae
Family Ransomnematidae
Family Lepidonematidae
Family Hethidae
Family Carnoyidae
Family Lauronematidae
Family Rhigonematidae
Family Blattophilidae
Order Spirurida (animal parasites)
Suborder Camallanina
Superfamily Camallanoidea
Family Cucullanidae
Family Camallanidae
Family Anguillicolidae
Superfamily Dracunculoidea
Family Dracunculidae
Suborder Spirurina
Superfamily Spiruroidea
Family Spiruridae
Family Hedruridae
Family Tetrameridae
Superfamily Physalopteroidea
Family Physalopteridae
Superfamily Filarioidea
Family Desmidocercidae
Family Filariidae
Family Setariidae
Family Onchocercidae
Superfamily Gnathostomatoidea
Family Gnathostomatidae
Superfamily Rictularioidea
Family Rictularidae

Higher Classification of the Phylum Nematoda (continued)

Superfamily Thelazioidea
Family Thelaziidae
Superfamily Habronematoidea
Family Habronematidae
Superfamily Acuarioidea
Family Acuariidae
Superfamily Aproctoidea
Family Aproctiidae
Superfamily Diplotriaenoidea
Family Diplotriaenidae

Chapter 3

KEY TO ENTOMOGENOUS NEMATODES

The following key is designed for the identification of all nematodes that are associated with insects. A portion of these are considered biological control candidates and are discussed in the present work. The others, which are found in insects but do little consistent damage, are treated in an earlier work.[1]

The heteroxenous nematode parasites that utilize insects as intermediate hosts may produce disease and even kill their insect host, but because the completion of their cycle requires a vertebrate host, they are simply not practical candidates for the biological control of insects and will not be discussed further here.

Key to the Orders and Families of Nematode Parasites of Insects*,**

1. Heteroxenous parasites that utilize insects as intermediate hosts; only juvenile stages present in the host (rarely juveniles of some forms [e. g., *Rhabdochona* spp.] may mature to the adult stage in insect hosts); nematodes usually surrounded by host tissue or within host cells: rarely free in the intestinal tract; stylet rarely present and pharynx usually lacking valves — 2

1. Monoxenous parasites that utilize insects as the sole or definitive host; either juvenile stages alone or juvenile and adult stages in the host; nematodes usually free in the gut or body cavity, not surrounded by host tissue; stylet may be present, especially in the invasive stages; if stylet is absent, then pharynx usually contains a valve (may be reduced) in the median or basal bulb; if stylet and pharyngeal valves are both absent, then the pharynx is composed of a thin cuticular tube or muscular tissue (no glandular portion) and juveniles or juveniles and adults, respectively, occur in the host — 3

2. Pharynx composed of a muscular and glandular portion; rectal cells usually prominent; basal portion of pharynx of fully developed juveniles (L-3) without valve — Spirurida

2. Pharynx composed of only muscular tissue; rectal cells absent or reduced; if prominent rectal cells are present, then the basal portion of the pharynx of third-stage juveniles is valvated — Ascaridida — 6

3. Stylet usually present, at least in the infective stages; juvenile or juvenile and adult nematodes occur in the body cavity (rarely in the gut) of living hosts; if stylet is absent, then adult stages also occur in the host — 4

3. Stylet absent; only juvenile nematodes occur in the body cavity of living hosts; juveniles and adults may occur in the gut of living hosts or body cavity of dead insects — 5

---

* Nematodes which have a simple phoretic association with insects or use insects as transport or paratenic hosts are not included here.

** See Poinar[2] for an illustrated version of this key.

4. Pharynx composed of a long narrow tube associated with a stichosome (best seen in early parasitic stages) or a tetrad of cells; stage which leaves host is usually elongate and threadlike; testes paired; stylet present in infective stage juveniles — Enoplida (Mermithoidea) — 7

4. Pharynx not composed of a long narrow tube associated with a stichosome or tetrad of cells; stage which leaves host never elongate and threadlike; testis single; stylet present in infective stage females — Tylenchida — 8

5. Adult and juvenile nematodes occur in the gut of living insects; nematodes with a valvated basal bulb — Oxyurida

5. Adult and/or juvenile nematodes occur in the body cavity of living or dead insects; only juveniles normally occur in the gut; if both juveniles and adults occur in the digestive tract, then the median bulb only is valvated — Rhabditida — 12

6. Valvated basal bulb present in mature (L-3) juveniles — Subuluroidea (Subuluridae)

6. Valvated basal bulb absent in juvenile stages — Seruatoidea (Seuratidae)

7. Nematodes threadlike; generally over 1 cm in length; juveniles (rarely adults) only present in hosts — Mermithidae

7. Nematodes not threadlike; generally under 1 cm in length; juveniles and adults occur in the host; mating may occur and eggs may be deposited, but they do not hatch in the living host — Tetradonematidae

8. Median pharyngeal bulb containing the outlet of the dorsal pharyngeal gland; three pharyngeal glands present — Aphelenchoidea — 9

8. Median pharyngeal bulb absent or, if present, not containing the outlet of the dorsal pharyngeal gland; two or three pharyngeal glands present — 10

9. Mature parasitic females deposit eggs in body cavity of host — Entaphelenchidae

9. Only juvenile stages associated with the host; may occur in the malpighian tubules, trachea, intestine, or body cavity — Aphelenchoididae

10. Free-living stages with only two pharyngeal glands with their outlets occurring in the middle of the pharyngeal tube; large swollen parasitic females with an everted reproductive system or, if not everted, then the juvenile nematodes are capable of reaching maturity and reproducing within the mother nematode — Sphaerularioidea (Sphaerulariidae)

10. Free-living stages with three pharyngeal glands, the dorsal outlet opening beneath the stylet; large swollen parasitic females without everted reproductive system (except in *Sphaerulariopsis,* a bark beetle parasite) — Neotylenchoidea — 11

11. With two types of free-living females, one capable of establishing a plant feeding cycle and the other capable of parasitizing insects — Neotylenchidae

11. With one type of free-living female which is only capable of infecting insects; no plant feeding generation established — Allantonematidae

12. Pharynx with a valve in the median bulb; basal bulb glandular, nonvalvated; occur in the gut or body cavity of insects — Diplogasteridae

12. Pharynx with a valve (sometimes vestigial) in the basal portion — 13

13. All stages occurring in the intestinal tract of living insects; valve in basal bulb usually reduced — 14

13. Usually in body cavity of living or dead insect; at most, only juveniles present in intestinal tract of living insect; valve in basal bulb not reduced — 15

14. Pharynx without a basal bulb; occurs in the intestinal tract of adult syrphid flies — Syrphonematidae (contains the single species *Syrphonema intestinalis* Laumond and Lyon)

14. Pharynx with a basal bulb; occurs in the intestinal tract of beetles — Rhabditidae (*Eudronema intestinalis* Remillet and van Waerebeke)

15. Adults and juveniles found in living carabid beetles; males with a bursa supported by papillae swollen at tip — Carabonematidae (contains the single species *Carabonema hasei* Stammer and Wachek)

15. Not found in living carabid beetles; if bursate, then not supported by papillae swollen at tip — 16

16. Valve in basal bulb reduced; reproduce in insects killed by a specific bacterium introduced with the infective stage — 17

16. Valve in basal bulb distinct, reproduce in living or dead insects; sometimes only juveniles occur in the host — 18

17. Males with a bursa; tail of reproducing female long and pointed — Heterorhabditidae (contains the single genus *Heterorhabditis* Poinar)

17. Males without a bursa; tail of reproducing female short and blunt — Steinernematidae

18. Juveniles occasionally found in the body cavity of insects; adult female has single ovary which extends back past the vulvar opening — Panagrolaimidae (representatives found in insects belong to the genus *Panagrolaimus*)

18. Juveniles or adults and juveniles occurring in various locations in insects (body cavity, malpighian tubules, head glands, colleterial glands, bursa copulatrix, uterus, aedeagus) — Rhabditidae

Chapter 4

NEMATODE GROUPS

I. MERMITHIDAE

The mermithids constitute a very important group of obligately parasitic inverte-brate nematodes. They are widely distributed in both aquatic and terrestrial habitats and attack a wide range of invertebrates. Most of their hosts are insects, but spiders, crustaceans, earthworms, leeches, and molluscs may also be included. In most cases, the hosts die soon after the mermithids complete their development and enter the en-vironment. Herein lies the importance of this group of nematodes; essentially, almost every mermithid is a potential biological control agent. However, most mermithids have never been reared completely through their life cycle in the laboratory and thus are not included here.

The mermithid life cycle can be somewhat complex when a second paratenic host is involved. However, most mermithids investigated thus far have a general life cycle involving only one insect or invertebrate host. An infective stage juvenile mermithid (all shown to be second stage juveniles thus far) emerges from an egg and actively seeks out a potential host. Once contact is made, the nematode begins penetrating through the integument of the insect with the aid of a stylet and enzymatic secretions.

After entering the host's hemocoel, the nematode initiates development, generally resting freely in the body cavity but sometimes entering nerve tissue for an initial period of adjustment. After completing its period of growth and development as a parasitic juvenile, the mermithid leaves the host, generally by boring through the integument. After entering the environment, the postparasitic juveniles molt to the adult stage, then mate (unless they are hermaphroditic or parthenogenetic), and the females deposit their eggs.

There are a few exceptions to this general pattern. Some mermithids (e.g., *Hydrom-ermis contorta* (on Linstow) and *Capitomermis crassiderma* Rubt.) molt to the adult stage while still inside their hosts. Four molts have been shown to occur during the development of *Romanomermis culicivorax* and probably occur in other mermithids also.

Another exception to the above pattern is shown by *Mermis nigrescens,* a parasite of Orthoptera and other insects. During rainstorms, the gravid females of this mermi-thid emerge from the soil and ascend vegetation to deposit their eggs. These eggs are drought resistant and remain on the leaf until it drops or is eaten by a potential host. The eggs hatch in the intestine of the insect, and the infective stage juvenile penetrates into the hemocoel and initiates development. The remainder of the life cycle is similar to that mentioned above. A rather unusual and recently discovered exception to the general life cycle pattern is the presence of paratenic hosts in the development of mer-mithids.

It has been shown that *Pheromermis pachysoma* Poinar, Lane, and Thomas and probably *P. myopis* Poinar and Lane are dependent on another invertebrate to reach their final host. Both of the above mermithids attack predatory insects — yellowjackets and horseflies, respectively. By entering the tissues of the prey, the infective stage mer-mithids can then gain entrance to the predatory host when the latter is either inaccess-able or has too thick a cuticle for penetration.

The most successful mermithid found thus far, from the standpoint of biological control, is *R. culicivorax.* This nematode has been mass produced and used as a bio-logical control agent in various parts of the world. This program serves as a good example of what can be done with a promising nematode parasite of insects.

Individual mermithid species will be discussed under the group of insects they parasitize. An up-to-date key is presented to point out important taxonomic differences between the known mermithid genera. (In the key, synonyms are enclosed in parentheses.) Over 200 species of mermithids have been described and many more still await characterization.

Key to the Genera of the Mermithidae[2]

1. Postparasitic juveniles and adults with two lateral lip papillae and four cephalic papillae; postparasitic juveniles lacking a tail appendage — *Mermis* Dujardin
1. Postparasitic juveniles and adults not with the above head papillae arrangement; postparasitic juveniles usually with a tail appendage; rarely a scar (as in *Agamermis*) or nothing (as in *Pheromermis* and some species of *Mesomermis*) — 2
2. Postparasitic juveniles and adults with two lateral lip papillae and six cephalic papillae — *Neomermis* v. Linstow (*Octomermis* Steiner)
2. Postparasitic juveniles and adults without lip papillae — 3
3. With only two cephalic papillae — *Orthomermis* Poinar
3. With four or six cephalic papillae — 4
4. With only four cephalic papillae — 5
4. With six cephalic papillae — 7
5. Vagina straight, barrel-shaped; adult cuticle lacking cross fibers (best seen at head or tail end); found in aquatic insects — *Pseudomermis* de Man (*Tetramermis* Steiner)
5. Vagina V- or S-shaped; adult cuticle with cross fibers; found in terrestrial insects — 6
6. Vagina bent in transverse plane to line of body; eight hypodermal cords; eggs covered with minute processes — *Allomermis* Steiner (*Melolonthinimermis* Artykhovsky)
6. Vagina parallel to line of body; six hypodermal cords; eggs not covered with minute processes — *Pheromermis* Poinar, Lane, and Thomas
7. Adult cuticle with cross fibers; generally parasites of terrestrial insects — 8
7. Adult cuticle lacking cross fibers; generally parasites of aquatic insects — 17
8. Males absent; vagina straight, barrel-shaped; vulva flanked by two wide lips — *Tunicamermis* Schuurmans-Stekhoven and Mawson
8. Males present; vulva not surrounded by two wide lips — 9
9. Male with single spicule; parasites of aguatic insects — 10
9. Male with paired spicules; parasites of terrestrial and aquatic insects — 11
10. With six hypodermal cords — *Paramermis* v. Linstow
10. With eight hypodermal cords — *Eumermis* Daday
11. Spicules medium to long; more than two times tail diameter — 12
11. Spicules short; less than two times tail diameter — 14
12. Spicules twisted; vagina S-shaped — *Amphimermis* Kaburaki and Imamura (*Complexomermis* Filipjev)
12. Spicules parallel; vagina straight, barrel-shaped — 13
13. Spicules short to medium in length, generally from two to four times tail diameter; in aquatic insects — *Bathymermis* Daday
13. Spicules long, usually more than four times tail diameter; in terrestrial insects — *Skrjabinomermis* Polozhentsev
14. Amphids cup-shaped, medium to large; found in aquatic insects — 15
14. Amphids thread-like, small; found in terrestrial insects — 16
15. Vagina straight, barrel-shaped; ovoviviparous — *Heleidomermis* Rubtzov
15. Vagina S-shaped; oviparous — *Amphidomermis* Filipjev

16. Preparasitic juveniles amputate tail upon entrance into the host; postparasitic juveniles with a scar on the tail — *Agamermis* Cobb, Steiner, and Christie

16. Preparasitic juveniles do not amputate tail upon entrance into the host; postparasitic juveniles with a tail appendage-*Hexamermis* Steiner (*Amphibiomermis* Artyukhovsky) (*Oesophagomermis* Artyukhovsky)

17. Males with a single spicule or spicules fused — 18

17. Males with paired, parallel, separate spicules — 23

18. Spicule medium to long, more than two times tail diameter — 19

18. Spicule short, less than two times tail diameter — 21

19. Two hypodermal cords; mouth terminal — *Phreatomermis* Coman

19. Six or eight hypodermal cords; mouth terminal or shifted ventrally — 20

20. Mouth terminal; spicule J-shaped; six hypodermal cords; vulval flap present — *Lanceimermis* Artyukhovsky

20. Mouth usually shifted ventrally; spicule curved but not J-shaped; six or eight hypodermal cords; vulval flap absent — *Gastromermis* Micoletzky

21. Amphids small; vulva protruding; spicule shorter than body width at anus — *Perutilimermis* Nickle

21. Amphids medium to large; vulva not protruding; spicule longer than body width at anus — 22

22. Eight hypodermal cords; pointed tail — *Hydromermis* Corti

22. Six hypodermal cords; blunt tail — *Limnomermis* Daday (*Comanimermis* Artyukhovsky)

23. Vagina straight, barrel-shaped — 24

23. Vagina S-shaped or elongate (extended or with a slight curve) — 26

24. Eight hypodermal cords — 25

24. Six hypodermal cords — *Mesomermis* Daday (*Psammomermis* Polozhentsev) (*Abathymermis* Rubtzov) (*Pologenzevimermis* Kirjanova, Karavaeva, and Romanenko) (*Neomesomermis* Nickle)

25. Spicules medium, from two to four times tail diameter — *Romanomermis* Coman (*Reesimermis* Tsai and Grundmann)

25. Spicules short, less than two times tail diameter — *Octomyomermis* Johnston (*Capitomermis* Rubtzov)

26. Four hypodermal cords — *Quadrimermis* Coman

26. Six or eight hypodermal cords — 27

27. Eight hypodermal cords — 30

27. Six hypodermal cords — 28

28. Spicules very long(more than ten times body width at anus) — *Drilomermis* Poinar and Petersen

28. Spicules, if males are present, short to medium (from two to four times body width at anus) — 29

29. Males rare or absent, females parthenogenetic; parasitize terrestrial insects — *Filipjevimermis* Polozhentsev and Artyukhovsky

29. Males present; parasitize aquatic insects — *Strelkovimermis* Rubstov (*Kurshymermis* Zahidov and Poinar) (*Diximermis* Nickle)

30. Vagina elongate, nearly straight; spicules small, shorter than tail diameter — *Culicimermis* Rubtsov and Isaeva (1975)

30. Vagina S-shaped, curved; spicules range from shorter than tail diameter to about nine times tail diameter — 31

31. Spicule length equal to or shorter than tail diameter; amphids small; cephalic crown well-developed — *Empidomermis* Poinar

31. Spicule length between 1 and 9 times tail diameter; amphids medium-large; cephalic crown poorly developed or absent — *Isomermis* Coman

A. Mermithids Infecting Aquatic Insects

Mermithid nematodes appear to be as abundant in an aquatic environment as a terrestrial one and, indeed, it may be impossible to find a nonpolluted body of fresh water that does not have an abundant population of mermithid nematodes.

Since the larval stages of several groups of medically important insects, such as mosquitoes and blackflies, develop in fresh water, their mermithid parasites have received a fair amount of attention, especially in the past 5 years. Thanks to the efforts of J. Petersen and his colleagues, the mosquito parasite *Romanomermis culicivorax* has been mass reared in vivo and distributed to many laboratories throughout the world. Not only has this resulted in many biological control trials, but it has supplied adequate material for basic studies on mermithid growth, physiology, nutrition, and host-parasite relationship.

In general, aquatic mermithids are easier to handle and study than terrestrial ones, mainly because the stages are easier to follow in water than in soil, and the hosts often have a shorter developmental period.

B. Mermithids Infecting Mosquitoes
1. The genus Romanomermis Coman 1961
There are currently five species of *Romanomermis* which are parasites of mosquitoes. At the present time, *R. culicivorax* is the most important since it has been mass produced and proven to be an effective biological control agent.

a. Description[5]
Romanomermis Coman is characterized by the presence of six cephalic papillae, eight hypodermal cords, a straight, barrel-shaped vagina, paired, separate spicules, with a length of two to four times the anal body width, and a cuticle lacking apparent cross fibers.

2. Romanomermis culicivorax Ross and Smith 1976 (syn. Romanomermis sp., syn. Reesimermis nielseni Tsai and Grundmann 1969 [in part])
When this nematode was first recovered, it was identified as a species of *Romanomermis* by Nickle.[3] Subsequently, Nickle decided it was identical to the then-called *Reesimermis neilseni* that had been described by Tsai and Grundmann from Wyoming in 1969 after he noticed that the spicules were not fused at the base as had been originally described by Tsai and Grundmann.[4] However, with this modification of the description of *Reesimermis*, Nickle failed to realize that this genus now was in no major way different from the earlier described genus *Romanomermis* Coman, 1961. In fact, for some unknown reason, Nickle[3] did not include the genus *Romanomermis* in his descriptions of select genera of the Mermithidae.

This matter was corrected by the welcome contribution of Ross and Smith.[5] These authors synonymized *Reesimermis* with *Romanomermis* and furthermore pointed out that what had been called *Reesimermis neilseni* in Louisiana was in actuality a different species from the Wyoming *Reesimermis nielseni*. The latter nematode was transferred to *Romanomermis*, and the former species was described as the new species *culicivorax* in the genus *Romanomermis*. The conclusions of Ross and Smith[5] were supported by findings by Petersen[6] which showed biological differences between *Romanomermis culicivorax* and *Reesimermis nielseni*.

a. Description
Adults — the mouth is terminal and amphids medium in size, without a commissure, slightly larger in the female. The amphidial cavity is variable and the pharyngeal tube is long, often extending to the posterior end of the trophosome. The tail tip bluntly

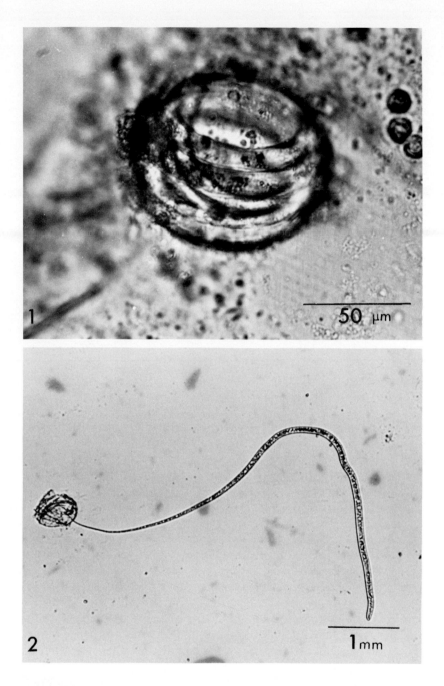

FIGURE 1 and 2. (1) Mature egg of *Romanomermis culicivorax* containing a preparasitic (infective stage) juvenile. (2) Preparasitic juvenile of *Romanomermis culicivorax* soon after emerging from an egg.

rounded. A reminant excretory pore is present at level with or slightly posterior to nerve ring. In females, the vulva protrudes slightly, the vulval flap is absent, and eggs are spherical (Figure 1). The male is shorter than the female. Spicules are long and narrow, gently curved, and tapering to the distal terminus. Genital papillae are in three rows, the middle row bifurcating around the cloacal opening. Quantitative values are given in Tables 1 and 2.

TABLE 1

Measurements of *Romanomermis culicivorax* Females[5]

Character	Value	
Length (mm)	16.8	(11.3—21.0)
Greatest width	191	(166—238)
Length amphidial pouch	14	(11—17)
Head to nerve ring	245	(203—282)
Length vagina	85	(74—100)
% vulva	48	(42—53)
Egg diameter	80	

TABLE 2

Measurements of *Romanomermis culicivorax* Males[5]

Character	Value	
Length (mm)	10.0	(7.2—12.9)
Greatest width	141	(105—156)
Length amphidial pouch	13	(12—13)
Head to nerve ring	214	(167—251)
Length spicules	397	(318—476)
Width spicules	5—8	
Length tail	140	(121—157)
Genital papillae	3 rows	

Infective stage juveniles (Figure 2) — Compared to other mermithids, the preparasitic or infective stages are very long (1000 to 1800μm). This stage contains a stylet, paired penetration glands, 16 stichocytes, an anus, and a long tail (see drawing in Poinar and Hess[7]). The genital primordium is located several body widths past the junction of the stichosome and intestine.

Parasitic juveniles (Figure 3) — The parasitic stages of *R. culicivorax* were described by Gordon et al.[8] (under the name *Reesimermis nielseni*). Depending on the age of the juvenile, the stichosome, trophosome, and genital rudiment can be found in various stages of development. The tail appendage becomes more pronounced with the age of the nematode. The cuticle of the parasitic stage is very thin and is easily ruptured upon handling. The trophosome gradually becomes darker and packed with storage material.

Postparasitic juveniles (Figure 4) — This stage is characterized by a well-developed tail appendage. As is the case with most mermithids, the third stage cuticle bearing the appendage and an inner, finer, fourth stage cuticle are shed simultaneously when the adult is formed.[9]

b. Bionomics

Life cycle — Petersen et al.[10] first reported on the life cycle of *Romanomermis culicivorax* (under the name of *Romanomermis* sp.) in the host *Psorophora confinnis*. Their observations showed that the infective stage juvenile mermithids entered the larval mosquito by cuticular penetration. Complete penetration into first instar larvae of *P. confinnis* occurred in less than 7 min. The infective stages died within 3 to 4 days if they did not find a host. At 24 to 26.5°C, the parasitic stage lasted about 8 days, and the nematodes then escaped through the cuticle of the host. The nematodes left third and fourth instar mosquito larvae and were not carried into the pupal or adult stages. The postparasitic juveniles molted to adults, mated, and deposited eggs in 11 to 13 days.

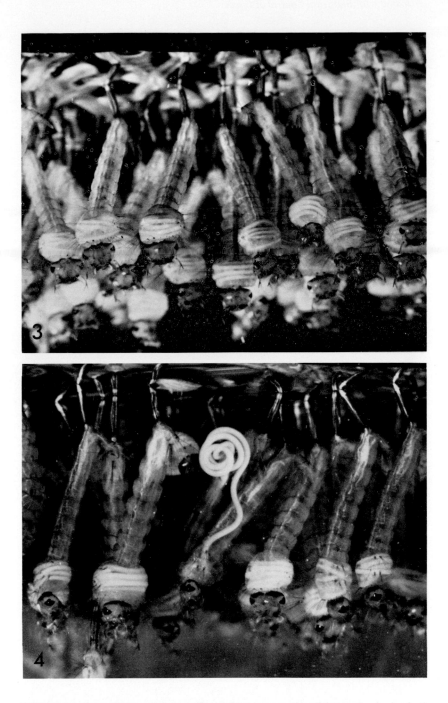

FIGURE 3 and 4.   (3) Parasitic juveniles of *Romanomermis culicivorax* developing in the thoraces of mosquito larvae (*Culex pipiens quinquefasciatus*). (Photo courtesy of J. J. Petersen, U.S. Department of Agriculture — Agricultural Research Service, Lake Charles, Louisiana.) (4) Postparasitic juvenile of *Romanomermis culicivorax* leaving its mosquito larva host (*Culex pipiens quinquefasciatus*). (Photo courtesy of J. J. Petersen, U.S. Department of Agriculture — Agricultural Research Service, Lake Charles, Louisiana.)

The eggs initiated development soon after oviposition and took at least 7 days to mature and hatch in water. The entire life cycle lasted about 4 weeks in the laboratory, and the adult females lived for more than 6 months after leaving the host. The present author noted that the preparasitic juveniles contained large amphidial openings (Figure 5) which are probably associated with host attraction.

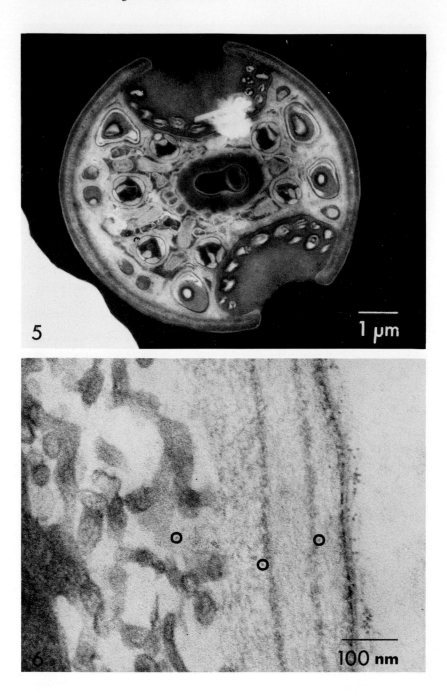

FIGURE 5 and 6.   (5) Electron micrograph through the amphidial openings of a preparasitic juvenile of *Romanomermis culicivorax.* (6) Electron micrograph of the body wall of
a 4-day-old parasitic juvenile of *Romanomermis culicivorax* incubated in ferritin for 4 hr.
Note ferritin particles on surface of nematode as well as inside the body (circles).

Further studies on the development of *R. culicivorax* were reported by Petersen[11] in
1975 (under the name *Reesimermis nielseni*). In mass cultures of this nematode, Petersen reported that about 10% of the males reached the adult stage 10 days after emergence, and it took 50 days for the entire male population to reach the adult stage. Of
those, 75% were still active 180 days after emergence. The females were about 10 to

14 days slower than the males, and all reached the adult stage only after 70 days. The first females oviposited 25 to 30 days after emergence. Petersen did not offer any reasons why these figures were so much longer than his original findings; however, it points out the increased variability one achieves with mass cultures of the nematodes. Since multiple parasitism was very common, it is possible that the smaller nematodes took longer to develop or perhaps crowding had an effect on maturation.

Petersen[11] also reported that infectivity of *R. culicivorax* preparasites at ambient temperature declined very rapidly after the first 24 hr. At 72 hr, there was little infectivity, although nearly 50% of the preparasites were still actively swimming.

Development — *R. culicivorax* develops fairly rapidly inside the hemocoel of mosquito larvae and is ready to emerge in 5 to 10 days, depending on the temperature. Since the first molt occurs in the egg, it is the second stage that initiates development in the mosquito. There is one molt inside the host, and the final two molts occur after the mermithid has left the mosquito and entered the environment.[9] Both Gordon et al.[8] and Otieno[12] noted an increase in growth rate on the fourth and third day of development, respectively. This sudden increase in growth was probably associated with the shedding of the cuticle.

Nutrients are taken in through the cuticle of the parasitic stages of *R. culicivorax*, and this is probably one of the reasons why growth is so rapid in this and other mermithids. Poinar and Hess[13] showed that the body surface of the parasitic stages of *R. culicivorax* was composed of three membranes overlying a layer of hypodermal cells. In all three membranes, there were pores ranging from 70 to 110 Å in diameter (Figure 6). These pores were large enough for the passage of ferritin particles and thus would allow passage of all nutrients required for growth. The gradient for this passage is probably established by the underlying hypodermal cells whose microvilli also take up ferritin particles.

Ecology — *R. culicivorax* naturally occurs in Louisiana and Florida and probably other southern states. It does not appear to be abundant in any one locality. After examining hundreds of locations in several areas in Louisiana, Petersen et al[10] found the mermithid in only five ponds. These ponds were all fresh water and located in semiopen areas of piney woods. The water never showed a salinity exceeding 96 ppm total soluble salts. Originally 13 species of mosquitoes were found to be natural hosts of *R. culicivorax*, and the percentage of parasitism ranged considerably (Table 3). The highest levels of parasitism occurred when the water level of pools fluctuated but never dried up completely. When water levels receded, larvae of the permanent water species

TABLE 3

Rate of Parasitism of *Romanomermis culicivorax* in Natural Mosquito Hosts in Louisiana[10]

Host	Percentage of parasitism observed in field
Aedes atlanticus Dyar and Knob	—
A. mitchellae (Dyar)	—
A. vexans (Mg.)	0—30
Anopheles crucians (Wied.)	0—30
A. punctipennis (Say)	0—30
A. quadrimaculatus Say	0—30
Culex erraticus (Dyar and Knob)	0—30
C. restuans Theobald	0—30
Psorophora ciliata Fab.	—
P. confinnis (Lynch-Arrib.)	0—93
P. discolor (Coq.)	—
Uranotaenia lowii Theobald	0—30
U. sapphirina (Osten-Sahen)	0—93

TABLE 7

Rates of Infection of *Culex tarsalis* and *Aedes dorsalis* to *Romanomermis culicivorax* at Different Temperatures[19]

Temperature (°C)	% of hosts infected[a]	No. nematodes per host[a]
	Culex tarsalis	
18	93.6[b]	2.59[b]
16	86.0[b]	1.72
14	42.2	1.24
12	31.9[b]	1.15[b]
10	1.5	1.00
	Aedes dorsalis	
18	73.1[b]	1.86
16	71.2[b]	1.79
14	49.6	1.52
12	48.6[b]	1.44[b]
10	1.6	1.00

[a]   15 replicates of 30 larvae each.
[b]   Level of significance between larvae of the two species is P < 0.01 or higher.

TABLE 8

Effect of pH on Infection of *Culex pipiens fatigans* by *Romanomermis culicivorax* Preparasites[22]

pH value	No. of surviving larvae (N = 200) exposed to 2000 preparasitic nematodes	% of survivors infected
5.7	0	—
6.2	99	6
6.7	124	84
7.2	148	100
7.7	128	24
8.2	71	14
8.7	0	—

enced the sex ratio were host species and host diet. In the latter case, starved hosts harboring a single nematode produced 92% males, whereas normally fed hosts produced only 13% males.

Effect of pH — Experiments conducted by Chen[22] indicated that the pH concentration was an important factor in the process of infection (Table 8). Best results with *Culex pipiens fatigans* serving as host occurred at the pH range 6.7 to 7.2.

Effect of pesticides — Studies by Mitchell et al.[23] showed that concentrations of Abate,® Dieldrin,® and Gama HCH® did not adversely affect the ability of *R. culicivorax* to parasitize larvae of *C. pipiens fatigans* (Table 9). Additional studies by Levy and Miller[24,25] indicated that 24-hr exposure to concentrations of Abate,®, Malathion,® Baytex,® Dursban,® Dimilin,® and Altosid® produced no noticeable effects on the viability and infectivity of *R. culicivorax* preparasites. However, some

TABLE 9

Effect of Certain Insecticides on the Preparasites of *Romanomermis culicivorax* and Larvae of its Host *Culex pipiens fatigans*[23]

Insecticide and concentration	No. of exposed larvae (N = 100) that survived	% of surviving larvae infected
Abate®		
Control	76	100
0.001	60	100
0.005	0	—
0.025	0	—
Dieldrin®		
Control	93	89
0.002	86	89
0.004	43	91
0.1	27	96
0.5	7	86
Gamma-HCH®		
Control	75	100
0.02	74	100
0.1	71	100

TABLE 10

Effects of Several Pesticides on the Viability and Infectivity of Preparasitic *Romanomermis culicivorax* to Second Instar Larvae of *Culex pipiens quinquefasciatus*[24,25]

Chemical	Concentration (ppm)	Percentage preparasite mortality (after 24 hr) (N = 2340) Test 1	Percentage host infection Test 1 (N = 155)	Test 2 (N = 100)
Abate®	0.001	1.8 (0—3.9)	82 (76 — 88)	68 (64 — 72)
Malathion®	0.100	1.3 (0—4.0)	76 (68 — 84)	72 (68 — 76)
Baytex®	0.004	0.8 (0—2.7)	52[a] (32 — 72)	—
	0.0035	—	—	54[a] (44 — 64)
	0.003	—	—	64 (56 — 72)
Dursban®	0.004	0.8 (0—2.3)[b]	14[a] (4 — 24)	—
	0.001	—	—	68 (60 — 76)
Dimilin®	0.005	1.1 (0—2.4)	76	82 (76 — 88)
Altosid®	0.005	1.1 (0—2.2)	76	90 (88 — 92)
Control	—	0.7 (0—2.0)	80 (76 — 84)	72 (64 — 80)

[a] Significantly lower than controls at 0.05 level.
[b] Preparasite movement very sluggish.

concentrations of Baytex® and Dursban® resulted in a lower rate of infection over the controls (Table 10).

Effect of salts — The effect of salts on the infectivity of the preparasites of *R. culicivorax* was determined by Platzer and Brown.[26,27] Preparasites and mosquito larvae were placed in 125 mℓ of salt solution (0 to 20,000 mg/ℓ) for 24 hr at 27°C. The mosquito larvae were then returned to tap water and the number of postparasites noted. The average lethal concentration for 50% (LC$_{50}$) in milligrams per liter for nematode infectivity was as follows (after Platzer and Brown[26]):

Salt	Mg/l for LC$_{50}$
Potassium	29
Sodium	24
Calcium	17
Sulfate	78
Chloride	68
Carbonate	51
Nitrate	36
Nitrite	17
Phosphate	12

The authors concluded that the nitrate and phosphate levels in feedlot runoff and fertilizer plant waste water would be detrimental to *R. culicivorax* infection. Petersen and Willis[15] showed that a salinity of 0.4 *M* NaCl was inhibitive to *R. culicivorax*, and therefore this nematode would have no value as a control agent of saltmarsh mosquitoes.

d. Host Range

R. culicivorax was recovered from 13 species of mosquitoes in the genera *Aedes, Anopheles, Culex, Psorophora,* and *Uranotaenia* in Louisiana[10] and also occurred in natural mosquito populations in Florida.[28] Species of mosquitoes that support the development of this mermithid are listed in Table 10a.

Because of the importance of *R. culicivorax*, tests have been made to see if it can enter and develop in a range of aquatic invertebrates as well as vertebrates (see the section on environmental impact). The infective stages can penetrate into some insects but cannot complete their development (Table 11). In most instances, the preparasitic juveniles are destroyed by a host reaction soon after entering and seldom have an opportunity to develop. In such cases, the host is usually not harmed unless multiple penetrations occur. The case of blackfly penetration by *R. culicivorax* is especially interesting. Blackfly larvae apparently have poor defense reactions and neither melanize nor encapsulate the infective stages of *R. culicivorax* after penetration. However, no one has been able to successfully rear this mermithid in blackfly larvae since the latter generally die immediately or within several days after penetration. Some development of the nematode was noted in *Simulium damnosum,* however, before the host died[29] (Figure 13).

Although *R. culicivorax* can develop in a range of mosquito species, it frequently is unable to complete its development in a few culicines, including *Aedes triseriatus, Culex salinarius, Psorophora ferox* and *C. territans*. The nematodes always were destroyed in the latter host. Kerdpibule et al[30] were unsuccessful in their attempts to infect first stage *Mansonia uniformis* with this nematode, but the reason was not stated. See the section on immune responses for further discussion of this subject.

A point might well be made here concerning the potential penetration hosts of *R. culicivorax*. Results of any infection experiments may depend on how the test is conducted, i. e., the size of the container, the number and age of the preparasitic juvenile nematodes, the stage of the experimental host, etc. This fact was made clear by Hansen and Hansen[29,31] during their work on the penetration of *R. culicivorax* into blackflies. They discovered that penetration into larvae of *Simulium damnosum* would not occur if conditions were unsuitable.

e. Culture

In vivo — Petersen et al.[10] made a very significant discovery when they noted that the embryonated eggs of *R. culicivorax* would remain for long periods in moist sand

TABLE 10a

Developmental Larval Mosquito Hosts of *Romanomermis culicivorax*

Host	Infection[a]	Ref.
Aedes aegypti (L.)	E	10
A. atlanticus Dyar and Knob	N	10
A. canadensis (Theobald)	E	10
A. dorsalis (Meigen)	E	19
A. fulvus pallens Ross	E	32
A. mitchellae (Dyar)	N	10
A. nigromaculis (Ludlow)	E	50
A. sierrensis (Ludlow)	E	10
A. sollicitans (Walker)	E	10
A. spencerii Theo.	E	33
A. sticticus (Mg.)	E	33
A. taeniorhynchus (Wied.)	E	10
A. thibaulti Dyar and Knob	E	10
A. tormentor Dyar and Knob	E	10
A. triseriatus (Say)	E	10
A. vexans (Meigen)	N	10
Anopheles albimanus Wied.	E	50
A. atropos Dyar and Knob	E	34
A. barberi Coq.	E	10
A. bradleyi King	E	10
A. crucians (Wied.)	N	10
A. franciscanus McCracken	E	35
A. freeborni Aitken	E	50
A. punctipennis (Say)	N	10
A. quadrimaculatus (Say)	N	10
Culex annulus Theobald	E	23
C. boharti Bohart	E	36
C. erraticus (Dyar and Knob)	N	10
C. erythothorax Dyar	E	36
C. fuscanus Wiedeman	E	23
C. fuscocephalus Theobald	E	23
C. nigripalpus Theobald	E	23
C. pallens Coq.	E	37
C. pipiens fatigans Wied.	E	23
C. pipiens molestus Forskal.	E	37
C. pipiens pipiens L.	E	50
C. pipiens quinquefasciatus Say	E	10
C. peccator Dyar and Knob	E	10
C. peus Speiser	E	36
C. restuans Theobald	N	10
C. rubithoracis (Leicester)	E	23
C. salinarius Coq.	E	10
C. tarsalis Coq.	E	50
C. tritaeniorhynchus summorosus Dyar	E	23
Culiseta incidens Thomson	E	36
C. inornata (Will.)	E	10
C. melanura (Coq.)	E	10
C. particeps (Adams)	E	36
Orthopodomyia signifera (Coq.)	E	10
Psorophora ciliata Fab.	N	10
P. columbiae Dyar	N	10
P. confinnis (Lynch-Ars.)	N	10
P. cyanescens (Coq.)	E	10
P. discolor (Coq.)	N	10
P. howardii Coq.	E	32
P. ferox (Hum.)	E	10
Uranotaenia lowii Theobald	N	10
U. sapphirinia (Osten-Saken)	N	10

[a] N, natural infection; E, experimental infection.

TABLE 11

Hosts into which *Romanomermis culicivorax* Can Penetrate, but which Rarely or Never Can Support Complete Development

Host family	Host	Stage[a]	Parasite development	Host response	Ref.
Chaoboridae	*Corethrella appendiculata* Grabham	L	No development	—	38
	C. brakeleyi (Coq.)	L	No development	No effect or death with multiple penetrations	38
Chironomidae	*Chironomus* sp.	L	No development	Melanization	38
Culicidae	*Aedes triseriatus* (Say)	L	No development	Melanization	36
		L	Some growth, rarely complete development	Melanization	39
	Anopheles sinensis Wied.	L	No development	Melanization	23
	Culex territans Walker	L	Little growth	Encapsulation	11
	Mansonia uniformis Theo.	L	Not reported	Melanization	30
	Psorophora ferox (Hum.)	L	Some growth, rarely complete development	—	39
Dytiscidae	Undetermined	L	No development	—	38
Haliplidae	Undetermined	A	No development	—	38
Hydrophilidae	Undetermined	L	No development	—	38
Simuliidae	*Simulium damnosum* Theo.	L	Initial development, death		29
	S. decorum Walk.	L	Not reported		40
	S. adersi Pomeroy	I	No development	Sluggish	31
	S. venustum Say	L	Not reported	—	40
	S. vittatum Zett.	L	Not reported	—	40

[a] L = larva; A = adult

and would hatch when the substrate was flooded. This was the basis for their establishment of a laboratory culture of this nematode. They first placed 30 to 40 postparasitic juveniles in round plastic pill boxes (3 cm in diameter) containing 5 to 7 mm of moistened sand. After capping the containers and holding them for 6 to 8 weeks, preparasitic juveniles could be obtained by flooding the container. In this period, the postparasitic juveniles had molted, mated, oviposited, and the eggs had embryonated (Figure 1).

An economic method for mass rearing *R. culicivorax* was developed later to meet the needs of extensive field testing.[41] The mosquito host chosen was *Culex pipiens quinquefasciatus* Say because it could be reared rapidly in crowded condtions, was easily maintained in colony, and was highly susceptible to the parasite. Nematode inoculum was obtained by flooding the nematode cultures in sand with a liter of chlorine free tap water, diluting and counting the nematodes in the suspensions, and then pouring them into paraffin-coated 136 × 52 × 5 cm galvanized trays filled with 21ℓ of tap water. At the same time, newly hatched mosquito larvae were added to make a ratio of 1:7.5 to 1:12 (host to parasites).

Since the infective stage juveniles are very sensitive to chlorine, 1 mℓ of 5% sodium thiosulfate (about 1 drop/ℓ) was added to each tray of water. For maximum production of female nematodes, a 1:12 ratio of host to parasite was used.

The nematodes began emerging from infected larvae about the same time the uninfected larvae pupated (Figure 7). At this stage the rearing trays were drained, and the mosquito larvae and pupae were collected on a 20-mesh screen, concentrated in smaller holding trays, and after 1½ hr were put into nematode-collecting containers. These containers each consisted of two 36 × 25 × 110 cm plastic trays designed to fit into each other with a clearance of 2 cm between the bottoms of the trays. The bottom of the upper container was replaced with 32-mesh nylon screen. Emerging postparasitic nematodes would sink, crawl through the screen, and collect in the bottom tray. Noninfected pupae were removed with ice water and used to maintain the mosquito colonies.

After washing, 10 to 15 gr of postparasitic juveniles were added to paraffin-coated aluminum cake pans filled to 1.5 cm with clean, coarse, sterile sand and covered with water to a depth of 1 cm. After about 3 weeks, the water was carefully decanted, and excess moisture was absorbed with paper towels. The cultures were stored and could be flooded anytime after 7 to 8 weeks to obtain infective stage juveniles (Figure 8).

However, higher hatches (1 to 10 × 10⁶ infective stages) occurred when the cultures were 11 to 16 weeks old. Additional hatches could be produced if the cultures were properly dried and stored for 3 to 4 more weeks. Sizable hatches were obtained through three floodings if the cultures were not more than 18 to 20 weeks old. Small hatches could be obtained after six floodings in cultures up to 34 weeks old.

Production costs (not counting equipment) amounted to about 7 to 10 cents per million infective stage nematodes. Using the rate of 1000 infective stages per square meter of water surface for field releases, treatment of a 2.5-acre surface would cost about $1.00.[41]

Later studies[42] showed that *R. culicivorax* was reared most rapidly and economically in 51 × 137 × 5 cm trays containing 20,000 hosts exposed at a ratio of 1:12 (host to parasites). The importance of host diet was also stressed. The optimum feeding schedule for highest production appeared to be 0.30, 0.45, and 0.60 gr of food per 1000 mosquito larvae on the first, second, and third days, respectively, and 0.90 gr each day thereafter. With these conditions, the yields were about 7.2 to 7.5 × 10⁵ female postparasites per 10⁶ hosts exposed.

In vitro — The usefulness of an in vitro culture of *R. culicivorax* for mass production under sterile conditions prompted several workers to initiate this phase of study.

FIGURE 7 and 8.   (7) Large numbers of postparasitic juveniles of *Romanomermis culi-civorax* after emergence from their hosts. (Photo courtesy of J. J. Petersen, U.S. Department of Agriculture — Agricultural Research Service, Lake Charles, Louisiana.) (8) Cultures of *Romanomermis culicivorax* ready for storage.

To date, however, normal development of *R. culicivorax* on artificial media has not been achieved.

Initial success was reported by Sanders et al.[43] when they achieved growth in commercially prepared Schneider's *Drosophila* medium (GIBCO) with 10% fetal calf serum added (GIBCO) in an atmospheric gas phase. The nematodes were grown in 5 mℓ of medium in a 50-nm diameter petri dish. Cultures were maintained at pH 6.5 to

7.0 at 25°C. The preparasitic juveniles were surface sterilized with antibiotics and antimycotics and placed into the media. Growth increased from 1 mm to 3 to 4 mm in length on the 15th day and 5 to 7 mm by the 25th day. Only a few nematodes reached 7 mm. The growth rate was calculated as one half to one third that of the nematode in vivo. Although the stichosome and trophosome became more pronounced, there was no gonadial development. Similar results were reported by Roberts and van Leuken,[44] who obtained nematodes approximately 6 to 7 mm in length within 6 weeks in tissue culture preparations.

The most recent attempt to culture *R. culicivorax* was made by Finney,[45] who used Grace's insect tissue culture medium with 10% fetal calf serum (GIBCO) inactivated for 30 min at 56°C. The medium, which was adjusted to 240 *M* osmolarity with distilled water and held at pH 6.4 to 6.5, was replenished every 3 to 4 days. After 6 weeks, the survival rate was estimated at 30%, and the length of the surviving nematodes was 12 to 14 mm; all became females with a vulval aperature and showed differentiation of the stichosome and trophosome.

f. Field Trials

The first field tests using *R. culicivorax* as a biological control agent were reported by Petersen and Willis.[46] The nematodes were obtained from in vivo laboratory cultures and the preparasitic nematodes were disseminated by using a 1-gal capacity compressed-air sprayer with a standard fan nozzle. Postparasitic nematodes were released in small groups (about 0.15 gm) randomly by hand. Mosquito larvae were sampled with standard pint dippers 1 to 3 days after treatment. The nematodes were applied to 11 sites known to be free of *R. culicivorax*. These sites were semipermanent or permanent ponds ranging from 209 to 3000 m² with a variable amount of vegetation. Most sites were exposed and contained high populations of mosquito predators, including top minnows.

The release of preparasitic juveniles (about 180 to 14,700 per square meter of surface area) resulted in 0 to 100% parasitism (Table 12). Thus, the effectiveness of this parasite as a biological control agent was demonstrated, and the nematode apparently became established at 7 of the 11 sites. The effect of dosage of preparasitic nematodes on the incidence of parasitism is presented in Table 13.

TABLE 12

Parasitism of Anopheles spp. in Ten Ponds in Louisiana After
Release of *Romanomermis culicivorax* Preparasites[46]

Site no.	No. preparasites released per m² surface area	No. mosquito larvae examined	% parasitism
1	5,000	42	67
2	14,700	67	79
3	2,400	61	77
4	300—1400	215	23—84
5	1500—3300	65	37—69
6	180—1000	116	18—86
7	230—740	56	14—71
8	670—1000	55	82—100
9	1300—3000	55	47—67
10	670—1000	118	20—42

TABLE 13

Effect of Different Dosages of *Romanomermis culicivorax* Preparasites on the Rate of *Anopheles* spp. Parasitism in Louisiana[46]

No. preparasites per m² surface area	No. treatments	% parasitism of second, third, and fourth instar *Anopheles* larvae
180—250	2	28
251—500	2	47
501—750	4	41
751—1000	4	64
1001—1500	3	63

Releases of preparasites of *R. culicivorax* were also made in rice fields in the California Central Valley in 1971.[47] The nematodes were released into rice fields by means of a 2-gal compressed-air hand sprayer. Mosquito larvae were sampled 1 to 2 days after treatment and once more within the following 5 days. One of the target species, *Culex tarsalis,* was not found parasitized whereas the other, *Anopheles freeborni,* indicated a 50 to 85% infection depending on the rate of preparasitic juveniles released. The results with *C. tarsalis* were considered inconclusive since the larvae were concentrated in a small strip that may not have been covered by the treatment. The encouraging results with *A. freeborni* suggested that further tests were worthwhile. The cost of producing the preparasitic nematodes was estimated at $0.50 per acre.

A third field test, again in rice fields but this time in Louisiana and directed against larvae of *Psorophora confinnis* and *A. quadrimaculatus,* was conducted in June 1972.[48] Because of low host densities, first stage larvae of *A. quadrimaculatus* were experimentally introduced into the test plots. The preparasites were applied in water from a knapsack sprayer at a pressure of about 40 psi and at a rate of 1 gal per treatment. The field applications were made 2 hr after the area was flooded to stimulate plant growth. Depending on the dose of preparasitics used, the percentage parasitism ranged from 6 to 38 in *P. confinnis* and 16 to 61 in *A. quadrimaculatus.*

Field experiments were also conducted against *Culex* mosquitoes in Taiwan.[23] *R. culicivorax* was cultured on a local strain of *C. pipiens fatigans,* and field releases were made either with a 3-gal compression sprayer fitted with a nozzle adjusted to deliver about 40 psi, a garden sprinkler, or by pouring a water suspension of the preparasites directly into selected pools. Two small pools containing larvae of *Culex tritaeniorhynchus summorosus* were treated with approximately 90,000 preparasites each. In addition, preparasites were used against the same host species in potholes and along the perimeter of paddy fields. No infected mosquito larvae were recovered in these experiments. However, another experiment where 116,000 preparasitic juveniles were placed in a small pond (1 × 4m) with a garden sprinkler yielded 11% parasitism of *C. tritaeniorhynchus summorosus.* Additional tests with *C. pipiens fatigans* as the host gave discouraging results. These low rates of infectivity were attributed to low water temperatures, predation by the ostracod, *Cyprinotus dentatis,* and extensive pollution in treated habitats.[49]

Certain insecticides did not adversely affect the ability of *R. culicivorax* to invade mosquito hosts (see Table 9).

Chapman[51] reported on field tests conducted in Thailand against larvae of *Culex pipiens fatigans* in polluted ditches and drains. Spraying *R. culicivorax* at doses exceeding 200,000 preparasites per square meter gave a low level of parasitism never exceeding 27%. No recycling of the parasite was noted.

Attempts to control *Anopheles* larvae in natural habitats in Louisiana were made

TABLE 14

Numbers of Parasitized *Anopheles crucians* Larvae from Sites Treated with
Romanomermis culicivorax in 1971[53]

	Mean percent parasitism			
Site	1971	1972	1973	1974
M-1	2	0	—	—
M-2	5	14	0	10
M-3	19	19	9	65
R-3	69	20	43	78
C-5	37	28	6	0

by Petersen and Willis.[52] Treated areas ranged from 225 to 550 yd² along the vegetated edges of 23 semipermanent or permanent ponds. Of these areas, 15 sites treated with about 1000 preparasitic juveniles per square yard gave 52 to 100% parasitism of *Anopheles* larvae. When 13 sites received 2000 preparasites per square yard, 36 to 100% of the mosquito larvae were infected.

To determine if *R. culicivorax* had become established at sites where it had been previously released, Petersen and Willis[53] returned to examine larvae of *A. crucians* at 5 sites treated in 1971, 5 treated both in 1971 and 1973, and at 12 sites treated in 1973. Four of the five sites treated in 1971 produced 7 to 52% parasitism of *A. crucians* larvae over the next 3 years (Table 14). Five of the six sites treated in both 1971 and 1973 produced from 2 to 51% parasitism during 1974; 5 of 12 sites treated in 1973 produced 11 to 85% parasitism in 1974. One site treated in 1973 produced 100% parasitism for 14 weeks and 94% parasitism during 1974. These results showed that *R. culicivorax* can become established and recycle itself in both permanent and semipermanent water sites.

Field tests of this nematode were conducted in Taiwan by Chen[22] against *C. pipiens fatigans*. Two breeding sites were chosen; the preparasitic nematodes were applied with a gravity type garden sprinkler or the postparasites were released directly into the sites. Up to 68% parasitism was achieved with the preparasitic nematodes, and recycling of the nematode occurred 10 weeks after treatment.

The results in the pond where the postparasitic juveniles had been released were negative. Apparently, conditions were not suitable for continued development of the nematode. In order to determine if the pH might have affected the mermithids, a test was made to determine the pH limits in which parasitism could occur (Table 8). It was interesting to see that the optimal value for parasitism was between 6.7 and 7.2, and no parasitism was observed at values lower than 5.7 or higher than 8.7.

Field tests conducted in three natural and two artificial ponds in California resulted in parasitism of *A. franciscanus, Culex tarsalis,* and *Culiseta inornata* (Table 15).[35] The preparasites were introduced into the water with an 8-ℓ sprayer, and the ponds were sampled 48 hr after treatment.

Releases of *R. culicivorax* were also made by the same authors in rice fields in California against *A. freeborni*. Up to 36% infection was achieved against natural populations of this host. The lower than expected rate of parasitism was attributed, in part, to differences in infectivity of the two nematode cultures used in the experiments.

Successful control of *A. freeborni* larvae was achieved with *R. culicivorax* at Searsville Lake in California. This lake was located in an ecological study site known as the Jasper Ridge Biological Preserve (Figure 9). Using a compressed-air sprayer with a standard fan nozzle to apply the preparasitic juveniles, Otieno[36] applied the parasites

TABLE 15

Rates of Parasitism by *Romanomermis culicivorax* of Three Mosquito Species
in Large Ponds in California[35]

No. preparasites per m²	Mosquito larvae	Parasitization		
		Anopheles franciscanus	*Culex tarsalis*	*Culiseta inornata*
0	Examined	5	31	29
	% Parasitized	0	0	0
1,000	Examined	6	30	27
	% Parasitized	67	13	22
10,000	Examined	12	35	36
	% Parasitized	83	57	56
25,000	Examined	2	26	38
	% Parasitized	100	62	58

at rates of 1200 per square meter of surface area. The treated areas varied from 15 to 149 m². Parasitism ranged from 70 to 97%, and infected mosquito larvae were recovered 54 days after release of the nematodes, indicating a recycling of the parasite (Tables 16 and 17). Parasitism was more successful in protected water filled with vegetation than in open water where wind may have caused nematode drift.

Although authors have shown the preparasitic juveniles of *R. culicivorax* sensitive to pH and salinity, Levy and Miller[54] surprisingly obtained parasitism of *Culex quinquefasciatus* larvae breeding in sewage settling compartments in Florida. Even with pH readings of 9.0, total alkalinity levels of 312 to 324 ppm and orthophosphate levels of 7.6 to 10.0 ppm in the release sites, 37.3 and 53.7% parasitism of *C. quinquefasciatus* was obtained. These results indicated the possible use of this mermithid as a control agent of mosquitoes in some polluted environments.

In an attempt to control mosquitoes breeding in a grassy field in Florida, Levy and Miller[55] released *R. culicivorax* at a rate of about 3.6×10^3 preparasites per square meter of water surface. The nematodes were applied to eight potholes and ditches with a compressed-air sprayer or directly from flasks. Results showed that 88 to 100% of the mosquito larvae sampled 24 hr after treatment harbored the parasite. The hosts included *Psorphora columbiae, P. ciliata, Culex nigripalpus,* and *Aedes taeniorhynchus.* The results also showed that natural populations of first and second instar larvae could be eliminated within 24 hr when the dosage was about 3.6×10^3 preparasites per square meter of water surface.

Aerial applications of *R. culicivorax* were made in Florida to control *Anopheles* and *Culex* mosquitoes by Levy et al.[56] (Figure 10). Levy et al.[57,58] had previously shown that an aerial spray system could be used to disseminate the preparasitic stages of this mermithid at 27 psi with no damage or loss of infectivity to the nematodes. Three ponds were sprayed via helicopter at the rate of approximately 3858 preparasitic juveniles per square meter of water surface. Mean levels of parasitism varied between 43 and 66% for *Anopheles* spp. and from 39 to 53% for *Culex erraticus.*

From the above field tests, there can be no doubt that under the right conditions, *R. culicivorax* is a very effective biological control agent of mosquitoes that could be used for short-term (inundative) or long-term (recycling) effects. We now know that some of the "right" conditions are a fairly warm temperature, the correct pH, little salinity and pollution, and the absence of nematode predators and probably pathogens. Although the preparasitic stages are resistant to low doses of some chemical pesticides, more data are necessary to determine the physical and chemical conditions of the environment that favor survival and parasitism by the nematode.

FIGURE 9 and 10. (9) Measuring out sand containing the mature eggs of *Romanomermis culicivorax* for application in Searsville Lake, California, against *Anopheles freeborni* larvae. (10) A Bell 47G helicopter with a Simplex low profile aerial spray system for applying *Romanomermis culicivorax* in a Florida pond. (Courtesy of Dr. Levy, Lee County Mosquito Control District, Fort Meyers, Florida.)

The Environmental Protection Agency determined that *R. culicivorax* was a parasite and therefore would not be subject to registration.[20] This mermithid was commercially produced by the Fairfax Biological Laboratories in Clinton Corners, New York. Their product, named "Skeeter Doom", was distributed in sand and available on request.

A second company, Nutrilite Corporation, in Lakeview, California is also producing

TABLE 16

Rates of Parasitism of *Anopheles freeborni* Larvae
by *Romanomermis culicivorax* at 2 to 54 Days After
Treating a 106-yd² Site at Searsville Lake, California[36]

Days after treatment	No. mosquito larvae collected	Percentage parasitism
2	85	71
7	129	78
13	194	81
19	219	80
26	228	80
33	236	81
40	240	80
48	242	81
54	244	81

TABLE 17

Rates of Parasitism of *Anopheles freeborni* Larvae
by *Romanomermis culicivorax* at 2 to 14 Days After
Treating a 178-yd² Site at Searsville Lake, California[36]

Days after treatment	No. mosquito larvae collected	Percentage parasitism
2	14	93
7	27	96
14	38	97

R. culicivorax for eventual sale. Since there can be a lot of variability involved in the
production of this nematode, and shipping conditions may be unfavorable, users of
this product should always examine a portion of the sample (place several cubic centi-
meters in water) before application to determine the actual number of viable preparasites.

Finney[40] and Hansen and Hansen[29] showed that the preparasitic stages of *R. culicivorax* could enter the early instars of blackfly larvae, and the latter authors conducted
preliminary field tests in the Ivory Coast, West Africa against larvae of *Simulium
damnosum*.[31] Vegetation bearing host eggs was placed in aluminum channels 2 m in
length. The preparasitic nematodes were introduced for 15 min as a siphoned suspension, and a 180 μm- mesh sac containing sand with mermithid eggs ready to hatch was
placed at the head of the channel (Figure 11). Simuliid larvae that became detached
from the substrate were collected up to 1 hr after application by an 85 μm- mesh screen
placed on the outlet of the channel. Parasitized first and second instar simuliid larvae
were recovered from the screen (Figures 12 and 13).

Although it is not yet known how far *R. culicivorax* can develop in simuliids, early
host instars are killed soon after penetration, and herein lies the possible use of this
nematode for control studies. Also, it was estimated that for every *S. damnosum* larva
penetrated by the nematode, approximately two others were mortally injured when the
preparasites attempted penetration.[31]

The field releases of *R. culicivorax* discussed in the preceding paragraphs are briefly
summarized in Table 18 as a convenience to the reader.

FIGURE 11. Introducing parasitic juveniles of *Romanomermis culicivorax* into an aluminum channel containing larvae of *Simulium damnosum* in the Ivory Coast, West Africa. (Photo by M. Laird.)

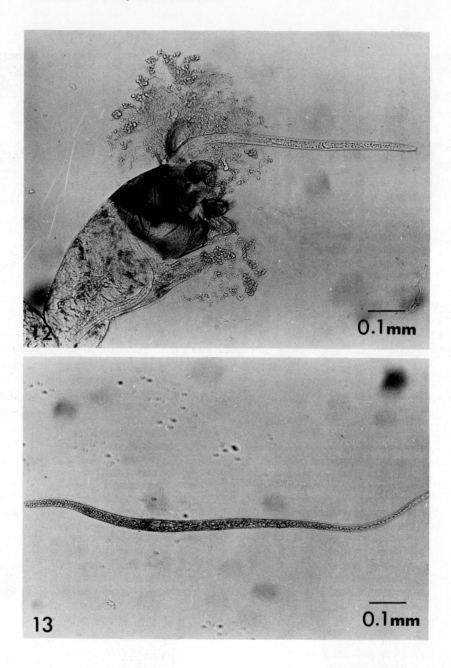

FIGURE 12 and 13.   (12) Preparasitic juvenile of *Romanomermis culicivorax* partly inside a first stage larva of *Simulium damnosum*. (Courtesy of E. Hansen.) (13) A juvenile of *Romanomermis culicivorax* that had remained 2 days in a second stage *Simulium damnosum* larva. Note initial increase in body width. (Courtesy of E. Hansen.)

3. *Octomyomermis* Johnson 1963

At least two species in the genus *Octomyomermis* are known to parasitize mosquitoes, *O. muspratti* and *O. troglodytis*.

a. Description

Members of the genus *Octomyomermis* have short spicules (less than twice anal body width), eight hypodermal cords, a straight barrel-shaped vagina, adult cuticle lacking apparent cross fibers, six cephalic papillae, and no lip papillae.

TABLE 18

Experimental Releases of *Romanomermis culicivorax* in the Field

Hosts	Area of treatment	Dosage	Results	Locality	Ref.
Anopheles spp. *Culex* spp. *Aedes* spp. *Psorophora* spp. *Uranotaenia* spp.	11 ponds	180—14,700 preparasites per m², clusters of postparasites	0—100% parasitism; recycling established	Louisiana	46, 55
C. tarsalis *Anopheles freeborni*	Paddies in two rice fields	500 and 1000 preparasites per yd²	0% parasitism of *C. tarsalis*; 50—85% parasitism of *A. freeborni*	California	47
Psorophora confinnis *A. quadrimaculatus*	Plots in rice field	180—1450 preparasites per yd²	6—38% parasitism of *P. confinnis*; 16—61% parasitism of *A. quadrimaculatus*	Louisiana	48
C. tritaeniorhynchus summorosus	Two pools, each about 1 m², potholes, and border of paddy field	90,000 preparasites in each pool; 11,500 nemas per pothole; 370,000 nemas around edge of paddy field	No infection	Taiwan	23
C. tritaeniorhynchus summorosus *C. pipiens fatigans*	Small pool (1 × 4 m) Ditches and drains	116,000 preparasites 200,000 preparasites per m²	11% infection Low levels of parasitism <27%	Taiwan Thailand	23 51

TABLE 18 (Continued)

Experimental Releases of *Romanomermis culicivorax* in the Field

Hosts	Area of treatment	Dosage	Results	Locality	Ref.
Anopheles spp.	Along edges of 23 ponds	1000 or 2000 preparasites per yd^2	36—100% of larvae infected, recycling established	Louisiana	52, 53
C. pipiens fatigans	Two ground pools; 5 m$^{2(A)}$ and 4.5 m$^{2(B)}$	(A) 1 × 10^4 preparasites; (B) 1000 postparasites	(A) Up to 68% parasitism, recycling; (B) no parasitism	Taiwain	22
A. freeborni	In ponds and along border of lake	1200 preparasites per m^2	70—97% parasitism; recycling evident	California	36
Simulium damnosum	Marahoue River	Siphoned suspension of preparasites and sand with embryonated eggs	First and second instar *S. damnosum* larvae were parasitized	Ivory Coast	31
C. pipiens quinquefasciatus	Two sewage settling tanks	6.4 × 10^4 and 1.1 × 10^5 preparasites per m^2	37 and 54% parasitism, respectively	Florida	54
Psorophora columbiae *P. ciliata* *C. nigripalpus* *Aedes taeniorhynchus* *Anopheles albimanus*	Eight potholes in grassy field	3.6 × 10^3 preparasites per m^2	96.5 ± 2.8% of all hosts parasitized	Florida	55
A. pipiens *pseudopunctipennis*	Lake Apastepeque	720 million preparasites over a 7-week period	Parasitism averaged 58 and 86%	El Salvador	59

TABLE 18 (Continued)

Experimental Releases of *Romanomermis culicivorax* in the Field

Hosts	Area of treatment	Dosage	Results	Locality	Ref.
1. *Anopheles franciscanus* 2. *Culex tarsalis* 3. *Culiseta inornata* 4. *A. freeborni*	Two artificial and three natural ponds	706—25,000 preparasites per m²	All species infected: 1. 67—100% parasitism, 2. 13—62% parasitism, 3. 22—58% parasitism, 4. 85% parasitism	California	35
1. *Anopheles quadrimaculatus* 2. *A. crucians* 3. *Culex erraticus*	Three ponds	3858 preparasites per m² of water surface	1 and 2. 43—66% parasitism, 3. 39—53% parasitism	Florida	56

TABLE 19

Description of the Males of *Octomyomermis muspratti* [61]

Character	Value
Length (mm)	7.6 ± 2.3
Diameter of amphidial opening	3—5
Head to nerve ring	250 (245—290)
Length spicules	115 ± 2.3
Genital papillae	Three rows
Length infective stage	800
Width infective stage	20

TABLE 20

Description of the Females of *Octomyomermis muspratti* [61]

Character	Value
Length (mm)	17.5 ± 5.3
Diameter of amphidial opening	3—5
Head to nerve ring	250 (245—290)
Egg diameter	84 × 74 (80—85—70—79)

4. Octomyomermis muspratti (Obiamiwe and MacDonald) 1973 (syn. Romanomermis sp., syn. Reesimermis muspratti Obiamiwe and MacDonald 1973)

a. Description

Adults — The head is narrowed at tip, sometimes set off from remainder of body by a slight constriction. Amphids are pear-shaped and larger in males. The mouth is terminal in cone-shaped depression. The pharyngeal tube extends to the posterior region of the body. In males, paired spicules are of equal length and separate. There are three rows of genital papillae, postanal genital papillae are indistinct and scattered. In females, the vagina is short, barrel-shaped, and symmetrical, the vulva is slightly protruding, and a vulval flap is absent. Quantitative measurements are given in Tables 19 and 20.

b. Bionomics[60,61]   Life cycle

— The eggs of *O. muspratti* hatch in 10 days to 2 weeks, and the preparasitic juveniles normally die within 24 hr if a host is not available. This stage enters the host by first wrapping its tail around the setae of the mosquito larva, then forcing its stylet back and forth against the host cuticle. Penetration is often through the anal papillae but may be through the body integument. Parasitic development takes 12 to 14 days at 25 to 27°C with a molt occurring on the eighth day. The postparasitic juveniles emerge from the mosquitoe's siphon or break through the abdominal wall of the larva or adult host. Eggs were deposited 2 to 3 weeks after the mermithids left their hosts.

Development — When first or second instar hosts were attacked, the nematodes completed their development and emerged through the wall of the later larval instars. If third or fourth instar larvae were attacked, the nematodes were carried over into the pupal and adult stages of the hosts.

When the host larvae were well-fed and the parasite numbers were high, female nematodes were mostly produced. When the diet was poor, the nematodes were smaller, and even when only one parasite was present, it was usually a male.

Using *Culex pipiens quinquefasciatus* as host, Petersen[62-64] studied *O. muspratti* in the laboratory from the standpoint of mass production and eventual field trials. He

TABLE 21

Hosts of *Octomyomermis muspratti*

Host	Stage	Infection[a]	Ref.
Aedes aegypti (L.)	L	N	65
A. calceatus Edw.	L	N	65
A. fulgens (Edw.)	L	N	65
A. haworthi Edw.	L	N	65
A. marshalli (Theo.)	L	N	65
A. metallicus (Edw.)	L	N	65
A. polynesiensis Mark	L	E	61
A. zethus De Meil	L	N	65
Anopheles albimanus Wied.	L	E	61
A. stephensi Liston	L	E	61
Culex nebulosus Theo.	L	N	63
C. pipiens molestus For.	L	E	61
C. pipiens fatigans Wied.	L	N	66
C. pipiens quinquefasciatus	L	E	62
C. torrentium Mart.	L	N	66

[a] N, naturally occurring infection; E, experimentally produced infection.

discovered that parasitic development took 11 to 14 days at 26 to 28°C and that about 37% of the hosts infected as third instars carried the nematode into the pupal stage. Second instar hosts were most suitable for infection and resulted in the highest yields of postparasitic juveniles. There were indications that *O. muspratti* is more tolerant of pollution and salinity and may be useful in certain environments restrictive to *R. culicivorax.*

Low male to female ratios were noticed while culturing this nematode, and it was shown that hosts with a single nematode produced fewer than 1% males, whereas hosts with eight mermithids produced about 40% males. However, about 60% of the females completed oviposition at male to female ratios of 1:20, and 100% completed oviposition at a 1:2 ratio. Thus, the low male to female ratio is compensated for by the ability of males to mate with a large number of female nematodes and remain reproductively active for at least 50 days after maturation.

c. Ecology[65]

This mermithid was originally collected from larvae of treehole breeding mosquitoes at Livingstone (3000-ft elevation) in Zambia (then Northern Rhodesia). Most observations were made on mosquitoes breeding in holes of the Mungongo tree (*Ricinodendron rautanenii*). Although the number of Mungongo trees supporting holes containing *R. muspratti* was only about 10%, the rate of parasitized mosquito larvae may be as high as 70 to 80%. A list of the hosts of this mermithid is presented in Table 21.

d. Culture

Muspratt[66] described a technique for culturing *O. muspratti* in the laboratory using *C. pipiens pipiens, C. pipiens fatigans,* and *C. torrentium* as hosts. As the mermithids emerged from their hosts, they were placed in a culture jar containing about 5 to 8 cm of sand on the bottom with 3 to 5 cm of water over the sand. After 3 to 4 weeks or longer, the standing water was withdrawn, the jar packed with absorbent cloth, and the lid tightened. Every 2 to 3 weeks over a period of 4 to 6 months, the cloth was removed, dried, and then replaced. This procedure was repeated until the sand became nearly dry; then the jar was set aside until infection was desired. At this time, water

TABLE 22

Measurements of *Octomyomermis troglodytis* Females[68]

Character (N = 10)	Value
Length (mm)	12.6 (10.2—15.0)
Greatest width	158 (120—165)
Head to amphidial opening	10 (7—10)
Length amphidial pouch	4 (3—7)
Diameter of amphidial opening	1.4 (0.6—2.6)
Head to nerve ring	236 (193—308)
Length vagina	83 (79—95)
% vulva	57 (54—64)
Egg diameter	81 (78—93)
Length tail appendage on post-parasitic juvenile	106 (91—109)

was added to the jar along with some second or early third instar mosquito larvae. In the meantime, the emerged postparasitic juveniles had molted and mated, and the deposited eggs had become embryonated. After water was added, hatching of the preparasites occurred over the following few days and infection was initiated.

Muspratt remarked that second or early third instar mosquito larvae should be used because first instars were often killed by the entering preparasitic juveniles. As can be seen, Muspratt's culture procedure is basically similar to the method used for rearing *Romanomermis culicivorax*. This nematode is now being cultured in Louisiana for eventual control studies.[62]

e. Field trials

Although Muspratt and others were successful in culturing this nematode, no mention is made of actual field releases of *O. muspratti* except by Laird.[67] These releases apparently were made by D. G. Reynolds against populations of *Aedes aegypti* and *C. pipiens fatigans* on the Pacific island of Nauru. Laird mentions that mermithid infections occurred in both of the above-mentioned hosts. Unfortunately, further information on these releases is not available at the present time. However, *O. muspratti* may be a successful biological control agent, especially when used against mosquito larvae breeding in treeholes under conditions unfavorable to the attack of other parasites.

5. Octomyomermis troglodytis Poinar and Sanders 1974

The above species of *Octomyomermis* was collected from larvae, pupae. and adults of the western treehole mosquito, *Aedes sierrensis* (Ludlow) in northern California. It was cultured for a short time in the laboratory and showed promise as a potential biological control agent.

a. Description[68]

Adults — The mouth is slightly displaced ventrally. Amphids are medium in size, without a commissure. The tail is bluntly rounded, without a mucron. The pharyngeal tube extends three fourths of body length. In the female, the vulva is slightly protruding and the vulvar flap absent; a rudimentary anal opening and abortive genital papillae are present. In the male, spicules paired, equal or subequal, and approximately equal in length to body width at the cloaca. Quantitative values of the adults are presented in Tables 22 and 23.

TABLE 23

Measurements of *Octomyomermis troglodytis* Males[68]

Character	Value
Length (mm)	8.2 (7.2—9.1)
Greatest width	117 (100—140)
Head to amphidial opening	9 (7—13)
Length amphidial pouch	5 (4—7)
Diameter of amphidial opening	1 (1—3)
Head to nerve ring	211 (193—254)
Length spicules	104 (83—130)
Width spicules	7 (6—9)
Length tail	145 (130—162)
Genital papillae	Three broken rows; each contains 19—23 preanal and 10—15 post-anal papillae
Length infective stage	(620—650)

Preparasitic juveniles — This stage contains a well-developed stylet, a pharyngeal tube, nerve ring, stichosome with 16 stichocytes, paired penetration glands, intestine, gonad primordium, anus, and elongate tail. The gonad primordium is located behind the junction of the pharynx and intestine.

Postparasitic juveniles — This stage molts twice after leaving the host to become an adult and contains a fairly long tail appendage.

Diagnosis — *O. muspratti* and *O. troglodytis* are the only two species in this genus which parasitize mosquitoes. It is interesting that both occur in treehole habitats. Diagnostic characters separating the two species are listed below.

Character	*O. muspratti*[61]	*O. troglodytis*[68]
Diameter of amphi dial openings (μm)	3.0—5.0	0.6—3.0
Mouth	Terminal	Displaced slightly ventrally
Pharyngeal tube	Extends almost to tip of body	Extends ¾ body length
Genital papillae	Median row longest	Lateral rows longest
Preparasitic juvenile (μm)	About 800 in length	620—650 in length

b. Bionomics[68]

Life cycle — Postparasitic juveniles generally emerged from fourth stage larvae of *A. sierrensis* (the western treehole mosquito), although pupae and adults were also infected. After entering the organic matter in the bottom of the treehole, the postparasites molted, mated, and oviposited. The final double molt occurred 10 to 18 days after emergence, and mating, and oviposition occurred 10 to 15 days after the final molt. The juveniles emerged from the egg about 15 days after oviposition. They were extremely active and most of the time remained near the surface of the water. Before penetration, the preparasite pressed its head against the host's cuticle and moved its stylet back and forth, during which time pharyngeal gland secretions were probably expelled through the stylet onto the insect's cuticle. A small hole was formed in the mosquito's cuticle in about 10 min, and the nematode suddenly straightened out and entered within a few seconds. Parasitic development was initiated at the base of the siphon, but after 8 to 10 days the nematode moved anteriorly (Figure 14). At 20°C, the entire developmental period lasted from 20 to 22 days. All hosts died soon after nematode emergence, and detailed examinations indicated fat body depletion.

FIGURE 14 and 15.   (14) An *Aedes sierrensis* larva containing a parasitic juvenile of *Octomyomermis troglodytis*. (15) Removing *Aedes sierrensis* larvae parasitized by *Octomyomermis troglodytis* from a treehole in California oak (*Quercus agrifolia*).

Field observations — Field studies were made in Marin County, California with *A. sierrensis* breeding in treeholes of California oak, *Quercus agrifolia*. Parasitized larvae could be found from January to June, depending on the rainfall, but the rate of parasitism was highest in late March, April, and early May when it reached 38% (Figure 15).

Since *A. sierrensis* may be a problem in urban areas near gardens and parks, *O. troglodytis* is a potential means of control for this and perhaps other treehole culicids.

c. Culture

O. troglodytis was reared through three generations in the laboratory on larvae of *A. sierrensis*; however, there is no present culture of this species available. It should not be too difficult to locate the parasite again since the host is widely distributed over parts of western America.

6. *Hydromermis churchillensis* Welch 1960

This is the only species of *Hydromermis* described from mosquitoes up to the present and was described from *Aedes communis* in Manitoba, Canada. Members of the genus *Hydromermis* Corti have the following characters: eight hypodermal cords, spicules fused or only one is present, cross fibers in the cuticle absent, six cephalic papillae, and an S-shaped vagina.

a. Description[69]

Adults — The mouth is terminal and the head slightly set off from the rest of the body by neck constriction. Cephalic papillae are in one plane. Amphids are pouch-shaped, connected by a commissure, with small, slitlike pores opening behind the lateral papillae. The pharyngeal tube extends past the vulva and the tail is pointed. In males, spicules are paired, fused for half their length, and blunt tipped.

TABLE 24

Measurements of Female *Hydromermis churchillensis* [69]

Character	Value
Length (mm)	16.7 (12.0—21.0)
Greatest width	116 (95—114)
Length amphidial pouch	8
Head to nerve ring	180—190
Length vagina	248
% vulva	T42 (42—52)
Egg diameter	45—52 × 32—45

TABLE 25

Measurements of Male *Hydromermis churchillensis* [69]

Character (N = 1)	Value
Length (mm)	11.9
Greatest width	130
Length amphidial pouch	9
Head to nerve ring	180—190
Length spicules	220
Width spicules	21

Preparasitic juveniles — This stage is 560 to 650 μm in length to 17 to 22 μm in width. The stylet is 9 to 11 μm long.

Postparasitic juveniles — This stage has a pronounced caudal appendage. Quantitative measurements of the adults are presented in Tables 24 and 25.

b. Bionomics[69]

Life cycle — Preparasitic juveniles enter a first or second stage mosquito larva, develop, and emerge from fourth instar hosts. If a host was not found, the preparasites died within 48 hr. The postparasitic juveniles emerged from natural openings or through intersegmental membranes. Rarely, the nematode was carried over into the adult stage of the host. The postparasites molted in 1 to 2 weeks after burrowing 5 to 15 cm into the sediment at the bottom of the pond. Mating occurred about 1 month after emergence, and oviposition began 1 to 2 weeks later. The females deposited clumps of 10 to 30 sticky eggs that adhered to the substrate. *A. communis* is considered a univoltine species that survives over winter in the egg stage.

Natural occurrence — Infected larvae of *A. communis* were found in forest pools and in transition forest pools, but not in the tundra zones. In those infected pools, the incidence of parasitism averaged from 5 to 15%. The parasite occurred in both permanent and temporary pools, although its random distribution could not be based on any special set of conditions. The parasite was recorded in Manitoba and Northwest Territories.

c. Host Range

Although *H. churchillensis* is primarily a parasite of *A. communis,* it has been recorded in several other hosts (Table 26).

d. Possible Application

This mermithid was apparently reared through its life cycle in the laboratory and

TABLE 26

Hosts of *Hydromermis churchillensis*

Family	Host	Stage	Infection[a]	Ref.
Chaoboridae	*Chaborus* sp.	L	N	69
	Mochlonyx sp.	P	N	69
Culicidae	*Aedes communis* (De G.)	L, A	N	69
	Aedes nearcticus Dyar	L	N	69
	Aedes nigripes (Zett.)	L	N	70

[a]   N, naturally occurring infection.

TABLE 27

Measurements of Male *Perutilimermis culicis* Stiles[3]

Character (N = 5)	Value
Length (mm)	8.0 (8.0—9.0)
Greatest width	180 (160—220)
Length spicules	90 (85—108)

could be considered a potential agent for use against mosquitoes in cold habitats. To the present author's knowledge, it has never been cultured on a large scale or applied in the field.

7. *Perutilimermis* Nickle[3] 1972

This genus was insufficiently described by Nickle,[3] but the species *Perutilimermis culicis* Stiles possesses one unique character that separates it from other mermithids. This is the single, short spicule of the male. The genus contains a sole species, *P. culicis* Stiles. From the partial description of *Agamomermis culicis* by Welch,[71] one could speculate that there are two different species under the name of *culicis* since Welch's and Nickle's descriptions are not the same.

The generic description is amended here by the present author on the basis of material sent to him by Dr. J. Petersen: mouth opening slightly ventral (Nickle described it as terminal); cuticle smooth; six head papillae; eight longitudinal cords; vagina long, S-shaped, vulvar flap absent; spicule single, short (length less than anal body width).

Unfortunately, adults were not available to the present author, so the specific description is based on the original, rather sparse, data.[3]

a. Description of P. culicis

Adults — Amphids are small and tail tips of both sexes are bluntly rounded. In the male, tail papillae are in three single rows. In the female, the vagina is long and S-shaped, the vulvar flap is absent, and the vulva is protruding slightly. The few measurements that are available on the adults are presented in Tables 27 and 28.

Preparasitic juveniles — This stage is about 709 μm long and 12μm wide.

Postparasitic juveniles — This stage contains a caudal appendage approximately 88 μm long.

b. Bionomics[72,73]

Life cycle — Egg development of *P. culicis* occurs within 28 to 48 hr after oviposition, and the embryos are active at 9 to 14 days after oviposition. The mature eggs

TABLE 28

Measurements of Female *Perutilimermis culicis* Stiles[3,72]

Character (N = 5)	Value
Length (mm)	13.0 (12.0—15.0)
Greatest width	255 (150—300)
% vulva	55 (52—62)
Egg diameter	45 (43—53)

hatch as soon as they are flooded with water, and the preparasitic juveniles remain passive for a short period (30 min or more) before becoming free-swimming. Some were able to penetrate into the hemocoel of mosquito larvae in less than 2 min. Although the preparasites can enter first instar host larvae, very little growth occurs until the host emerges as an adult. The parasites then may grow 30 mm in length in 7 to 9 days. Escape of the postparasitic juvenile usually occurs through the intersegmental membranes on either side of the eighth abdominal segment. After molting and mating, oviposition occurs from 4 to 21 days after nematode emergence.

Natural occurrence — Thus far, *Aedes sollicitans* is the only natural host of this mermithid. Of 2,362 specimens of *A. sollicitans* collected in southwestern Louisiana, 17% were infected with 1 to 23 nematode parasites. The rate of parasitism varied from 0 to 91% in different host populations. The parasites were recovered in Mississippi as well as Louisiana, and the original report of this nematode is based on material collected in New Jersey.[74]

Parasitism of *A. sollicitans* was widespread through the coastal marshes in Louisiana. Of 6094 adult hosts examined from 29 areas, 1026 contained one or more *Agamomermis culicis*. Indidividual populations varied from 0 to 96% in rate of infection. Seasonally, the highest rate of parasitism occurred from March to August, although infected hosts were recovered every month of the year. Precipitation, temperature, and tides were all considered as having an important influence on the extent of parasitism. The parasite was considered to have significantly reduced some populations of *Aedes sollicitans* in southwestern Louisiana.

c. Host Range[72,73]

A. sollicitans is the only known host of *P. culicis*. All attempts to develop *Agamomermis culicis* in other mosquito species failed. The preparasitic stages were observed to enter larvae of *Agamomermis taeniorhynchus, Culex pipiens quinquefasciatus,* and *Aedes aegypti,* however, the nematodes died before the hosts reached the adult stage. The present author is not aware of any field trials conducted with this species.

8. Strelkovimermis peterseni (Nickle) 1972

This mermithid was collected from *Anopheles* spp. in southwestern Louisiana. The description by Nickle[3] was incomplete, and it was difficult to determine the correct position of this mermithid. Thus, an examination of important characters omitted in the original description was made by the present author from material submitted by Dr. J. Petersen. It was discovered that the species *peterseni* possessed six head papillae, medium-sized amphids, six longitudinal cords, no apparent cross-striations in the cuticle, an S-shaped vagina, and two short, parallel spicules. The genus *Diximermis* therefore agrees in all basic characters with the description of the earlier genus *Strelkovimermis* Rubtsov, and the species *peterseni* is hereby transferred to that genus.

Characters of the species *S. peterseni* are as follows (based on Nickle[3] and observations made by the present author).

TABLE 29

Measurements of Female *Strelkovimermis peterseni*

	Value	
Character	After Poinar (present study)	After Nickle[3]
Length (mm)	14.6 (11.0—17.5)	12.0 (10.0—14.0)
Greatest width	185 (131—233)	170 (140—209)
Head to amphidial opening	23 (17—27)	—
Diameter of amphidial opening	7 (6—8)	—
Head to nerve ring	223 (200—246)	—
Length vagina	166 (124—186)	—
% vulva	58 (51—65)	55 (50—60)
Egg diameter	40 × 49 (37-41 × 46-53)	55 (52—58)

TABLE 30

Measurements of Male *Strelkovimermis peterseni*

	Value	
Character	After Poinar (present study)	After Nickle[3]
Length (mm)	8.9 (7.5—11.0)	8.0 (5.0—14.5)
Greatest width	115 (85—131)	280 (236—346)
Head to amphidial opening	19 (17—22)	—
Diameter of amphidial opening	6 (5—7)	—
Head to nerve ring	200 (193—223)	—
Length spicules	75 (62—78)	190 (182—213)
Anal body width	81 (65—93)	—
Length tail	99 (81—130)	—

a. Description

Adults — The mouth opening is shifted slightly to the ventral surface. Amphids are medium in size and shifted slightly to the dorsal side; the openings are slightly posterior to the cephalic papillae. In the female, cephalic papillae are usually more pronounced than in the male. The vagina is short and S-shaped. The vulval flap is absent (Nickle stated it was present but those specimens examined by the present author did not possess a vulval flap). The tail tip is rounded. In the male, there are two short spicules, equal or less in length to the anal body diameter, and three rows of genital papillae. The length of the spicules ranged between 62 and 78 μm on the specimens examined by the present author. Never did they reach 182 to 213 μm as described by Nickle.[3] Quantitative values are presented in Table 29 and 30.

b. Bionomics[75]

Life cycle — At 25°C, the eggs of *S. peterseni* hatched in about 14 days, and the preparasitic juveniles penetrated through the host's cuticle and began development. After 6 to 8 days, the postparasitic juvenile left the host by boring through the cuticle, generally in the anal region. Nematodes always emerged from the larval stage and were not found in the pupal and adult hosts. The postparasitic juveniles molted, mated, and deposited eggs in 9 to 12 days. Sex of the parasite was correlated with the infection rate, and as the number of nematodes per host increased, the ratio of males to females increased. When the hosts contained one and seven or more parasites, 5 and 91% males occurred, respectively.

TABLE 31

Host Range of *Strelkovimermis peterseni*[75]

Host	Stage	Infection[a]
Anopheles barberi Coq.	L	E
Anopheles bradleyi King	L	E
Anopheles crucians Wied.	L	N
Anopheles punctipennis (Say)	L	N
Anopheles quadrimaculatus Say	L	N

[a] E, experimentally induced infection; N, naturally occurring infection.

Natural occurrence[75] — Only three of many hundreds of sites sampled contained mosquitoes parasitized by *S. peterseni*; thus, it can be considered a rare species in southwestern Louisiana. In the field it was only collected from *Anopheles crucians* and *A. quadrimaculatus*. Parasitism of the former species ranged from 0 to 87%. Its rare occurrence may be associated with the observation that the nematodes were never carried over into the adult stage of the host. Thus, adult mosquitoes could not distribute the parasite.

c. Host Range[75]

Thus far, *S. peterseni* has only been recorded parasitizing species of mosquitoes in the genus *Anopheles* (Table 31). No development was reported in laboratory experiments with *Culex, Aedes,* or *Psorophora* species. Since penetration was not successful, the authors concluded that host resistance was due to some physical aspect of the cuticle, rather than a physiological factor.

d. Culture

S. peterseni has been maintained in the laboratory on *Anopheles quadrimaculatus* for at least 4 years.[75] During this period a gradual decline in nematode production was observed. Considering the possibility that *A. quadrimaculatus* had developed resistance to *S. peterseni*, Woodard and Fukuda[76] conducted some experiments and concluded that their hypothesis was correct. Their strain of *A. quadrimaculatus* which had been recycled for 4 years and was subject to selection pressure by the mermithid was compared with another strain that had not received nematode exposure.

Results showed that infection rates were almost twice as high in the unselected (previously nonexposed) strain as in the original strain. The authors concluded that the resistance was related to the activity of the mosquito larvae. The original strain contained larvae that moved vigorously and snapped at the attacking nematodes, whereas those larvae of the unselected strains were quieter. It was concluded that the practice of restocking the host colony with pupae that had escaped infection led to the development of this type of resistance. The present author has not seen any reports on the results of field trials with this mermithid species.

C. Mermithid Parasites of Midges (Chironomidae)

The following three mermithid species all attack midges of the family Chironomidae and have qualities marking them as possible biological control agents. A number of midges are host to mermithids, and it is probable that these nematodes constitute one of the most important groups of parasites of these hosts.

In discussing the prospects for biological control of chironomids in the Sudan, Wülker[85] noted that mermithid nematodes were the most common natural enemy of

TABLE 32

Measurements of Female *Hydromermis conopophaga*

| Character (N = 15) | Value | |
	Poinar[77]	Hominick and Welch[78]
Length (mm)	9.2 (6.0—13.0)	14.5 (11.5—15.9)
Greatest width	160 (110—230)	150 (120—170)
Head to amphidial opening	27 (22—34)	18 (13—23)
Length amphidial pouch	7 (6—9)	9-14 × 3-6
Diameter amphidial pore	—	3 × 4
Head to nerve ring	190 (150—230)	190 (140—240)
Length vagina	230 (190—300)	180 (160—230)
% vulva	47 (40—50)	47 (44—51)
Egg diameter	49 (46—52)	53-63 × 43-47

TABLE 33

Measurements of Male *Hydromermis conopophaga*

| Character (N = 15) | Value | |
	Poinar[77]	Hominick and Welch[78]
Length (mm)	6.1 (4.0—8.0)	9.6 (6.6—12.3)
Greatest width	130 (100—150)	103 (86—128)
Head to amphidial opening	24 (22—31)	19 (15—22)
Length amphidial pouch	16 (12—26)	8-12 × 3-8
Diameter of amphidial opening	5 (3—8)	3 × 4
Head to nerve ring	180 (170—230)	200 (160—220)
Length spicule	140 (110—167)	150 (120—170)
Width spicule	21 (16—34)	22 (16—27)
Length tail	150 (120—180)	140 (110—170)
Genital papillae	3 rows; 17—30	3 rows; 8—25 per row

the midges under investigation. He even mentioned the possibility of introducing exotic mermithids from Europe into the Sudan as possible biological control agents to augment populations of naturally occurring parasites. One mermithid parasitized at least four different species of midges in the Nile and probably would be a good biological control candidate.

1. *Hydromermis conopophaga* Poinar 1968

This nematode was recovered from midges of the genus *Tanytarsus* in artificial percolation ponds in California. It was reared through several generations in the laboratory, and field studies showed that it was a good biological control agent. A diagnosis of the genus *Hydromermis* is presented in Section I.B.6.

a. Description[77,78]

Adults — The mouth is displaced slightly ventrally and the opening is anterior to the head papillae. Amphids are oval, larger in males, and connected by a commissure. The tail varies from pointed to round. The excretory pore is 200 to 300 μm from the apex. The pharyngeal tube extends about one third of body length. In the female, a vulvar flap is present. In the male, the single spicule is curved, with minute granulations on the distal tip. Genital papillae often reach almost to the tip of the tail. Quantitative measurements of the adults are presented in Tables 32 and 33.

Postparasitic juveniles — This stage has a distinct tail appendage and is 54 to 110 μm long.

Preparasitic juveniles — This stage is 620 (560 to 680) μm long and is extremely active.

b. Bionomics

Life cycle — *Tanytarsus* hosts died 1 to 2 days after emergence of the postparasitic juveniles which then underwent their final molt to the adult stage about three days later (22°C). Mating immediately followed, and oviposition began about 8 to 9 days after emergence from the host. Hatching occurred 12 to 13 days after oviposition, and the emerging second stage juveniles entered the host by direct penetration through the cuticle. The actual process of entry took about 3 to 5 min. Penetration into later stage host larvae could account for the occurrence of nematodes in pupal and adult hosts. Parasitic development lasted from 7 to 11 days.

Incidence of infection — Chapman and Ecke[80] followed the population fluctuations of healthy and parasitized midges in percolation ponds in Santa Clara County, California over a 12-month period. Parasitized *Tanytarsus* larvae were recovered from November to July, although the highest rates of infection were found in spring and early summer, peaking at 70% infection in July 1967. The host populations fell soon after this, and the observed reduction appeared to be directly related to high rates of mermithid parasitism. Distribution of the mermithid naturally occurred through the infection of adult hosts, many of which exhibited intersexual morphological features. In California, several parasitic generations occurred throughout the year; however, Hominick and Welch[79] reported only a single generation of *H. conopophaga* at the latitude of Manitoba. This mermithid probably survived over winter in the egg stage and initiated new infections in the spring. The postparasitic juveniles left their hosts in June and July and molted in about 3 to 4 days, and gravid females were found in July and August. Intersexes of *Harnischia* sp. were found in individuals parasitized by *Hydromermis conopophaga*.

c. Host Range

In California, *H. conopophaga* was found parasitizing at least two species of *Tanytarsus* sp.[77] whereas in Manitoba, this nematode has been recorded from larvae, pupae, and adults of *Polypedilum simulans* Townes and *Harnischia* sp.[78] All hosts belong to the Chironomidae.

d. Culture

Tanytarsus midges could easily be maintained under laboratory conditions in containers of water containing small particles of ground rabbit chow. Infection readily occurred, and several generations were reared through their life cycle using this simple technique.

e. Application

To the author's knowledge, *H. conopophaga* has never been mass produced or released in the field.

2. *Limnomermis rosea* (Hagmeier) 1912 (syn. *Paramermis rosea* Hagmeier 1912, syn. *Gastromermis rosea* (Hagmeier) 1912)

This species was described by Hagmeier[81] from Neuhofen, Germany and has not been reported outside of Europe. It has been found exclusively parasitizing chironomid larvae. There is some question concerning its taxonomic position. However it has six longitudinal cords, and should be placed in the genus *Limnomermis* Daday.

TABLE 34		TABLE 35	
Measurements of Female *Limnomermis* *rosea*		**Measurements of Male** *Limnomermis rosea* [81]	
Character	Value	Character	Value
		Length (mm)	5.0—7.0
Length (mm)	6.0—9.0	Greatest width	106
Greatest width	110—140	Head to nerve ring	140
Head to nerve ring	160—180	Length spicules	80
% vulva	50	Length tail	96
Egg diameter	38	Genital papillae	3 rows of about 20 papillae each

a. Description[81]

Adults — The cuticle is without cross fibers. Amphids are large and located on the same level as the six head papillae, connected by a commissure. The mouth opening is shifted slightly ventrad. The pharyngeal tube extends about one half of the total length. The tail tip is bluntly pointed. In the male, there is a single short spicule. Quantitative measurements are given in Tables 34 and 35.

Preparasitic Juveniles[82] — This stage is 600 to 700 μm long and the greatest width is 10 μm.

b. Bionomics

Life cycle — Perhaps the most interesting aspect of the biology of this nematode is its manner of entry into the host which was described by Wülker.[83] He showed that the preparasitic juvenile injected a toxin into the host before and during penetration. This toxin paralyzed the host's muscles and thus allowed the nematode to enter the host much more easily. Penetration occurred in the posterior end of the paralyzed host through the mucous membrane of the hindgut. When two or more nematodes entered the host, there tended to be a greater number of males produced. Götz[82] noted that at 20 to 22°C, the eggs of *L. rosea* hatched in 10 to 12 days, and he followed the parasitic development of this species.

The postparasites of *L. rosea* are carried into the adult stage of the host and emerge only at this time. Parasitized adult midges swarm close to the water surface in a characteristic manner, which differs from the flight of unparasitized midges. Aside from inhibiting gonad development of the host, the parasite also causes very interesting morphological changes in adult midges, giving rise to intersex formation. These changes have been studied in detail by Wülker[84] and Götz.[82]

c. Host Range

Although Hagmeier[81] described *L. rosea* from adult nematodes collected from sand, subsequent workers recorded at least two chironomids as natural hosts of this species, and these along with experimental hosts are listed in Table 36. Even though very young host larvae are infected, the parasites always seem to emerge only after the host reaches the adult stage.

d. Culture

Development of *L. rosea* through several generations in the laboratory apparently presents no serious problems. Wülker[86,87] edited two films showing the penetration and development, respectively, of this mermithid in midges, and it should be possible to establish a breeding program if the need arises.

TABLE 36

Chironomid hosts of *Limnomermis rosea*

Host	Infection[a]	Ref.
Chironomus annularius Mg.	E	82
C. anthracinus Zett.	N	84
C. halophilus Kieff.	E	82
C. malanotus Str.	N	84
C. pseudothummi Str.	E	82
C. tentans Fab.	E	82
C. thummi thummi Kieff.	E	82
C. thummi piger Str.	E	82

[a] E, experimentally induced infections; N, naturally occurring infections.

e. Application

To date, the present author is not aware of any field releases or mass culture attempts with *L. rosea*.

3. Hydromermis contorta Kohn 1905

This midge parasite is one of the few mermithids that have been recovered from Europe and North America. It is one of the earliest mermithids to be described and attacks several genera of midges (Figure 16).

a. Description

Several workers have described *H. contorta* from various localities, and a summary of the quantitative measurements is presented in Table 37. The diagnostic features of this species are the pointed tails of both sexes, the characteristic bend of the spicule, and the biological feature that the final molts are completed in the host before emergence (Figure 17). Johnson[88] suggested that the European forms had a blunter head than the American *contortus,* but the curvature of the head appears to be a variable character. This species is not to be confused with the *contorta* described by von Linstow in 1889. The latter species possesses cross fibers and six hypodermal cords and belongs in the genus *Paramermis*.

b. Bionomics

Life cycle — The adults of *H. contorta* emerge directly from the larval host. Mating occurs at once, and oviposition begins 1 to 3 days after emergence. Depending on conditions, the eggs hatch in 7 to 10 days[90] or in 13 to 15 days.[88] Poinar and Tourenq[90] noted that it was possible to keep the preparasitic juveniles alive for 66 days at 5°C, although Johnson[88] gave a maximum survival time of 6 days at 17°C and 12 days at 12°C. Host infection occurred by direct penetration through the cuticle and parasitic development lasted 47 to 61 days.[88] As with other mermithids, the sex ratio is dependent on the number of parasites per host, with predominantly males occurring when five or more nematodes occur in a single host.[88] Whereas the parasite is bivoltene in lakes in North America,[88] it was regarded as univoltene in southern France.[90]

Ecology — The rate of infection of midges parasitized by *H. contorta* in southern France ranged from 0 to 46% with the highest rate occurring in April.[90] The rate of parasitism dropped off rapidly to almost 0% in June and July. Parasitism in the Camarque, in southern France where the water varied from fresh to marine, only occurred in fresh water. In North America, Johnson[88] recorded a monthly infection rate varying from 0 to 25% throughout the year. Highest infection rates were recorded in November, March, and April.

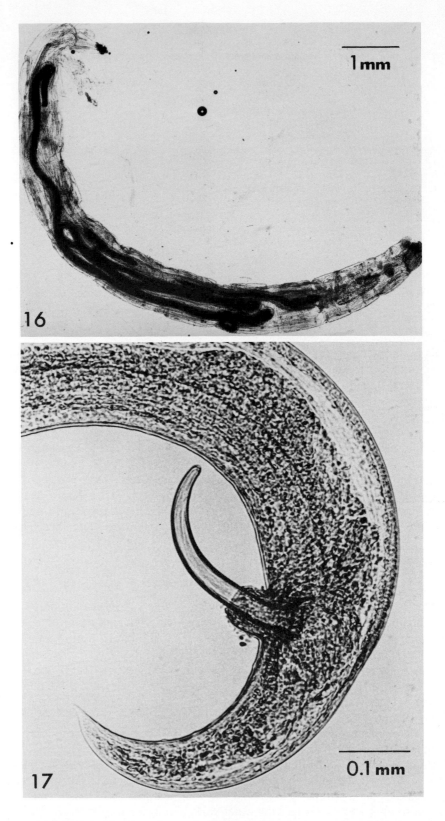

FIGURE 16 and 17 .   (16) Larva of the midge, *Chironomus annularius* parasitized by *Hydromermis contorta*. (17) Exserted spicule of *Hydromermis contorta*.

TABLE 37

Measurements of *Hydromermis contorta* (Kohn) From Various Localities

Character	Kohn[89]	Hagmeier[81]	Johnson[88]	Poinar and Tourenq[90]
Length (mm)				
Female	26—50	19—40	50 (23—70)	38 (31—54)
Male	13—26	14—25	26 (15—40)	23 (14—30)
Width				
Female	0.18—0.37	0.18—0.24	.31 (0.22—0.44)	0.27 (0.20—0.34)
Male	0.07—0.21	0.13—0.19	.24 (0.12—0.33)	0.17 (0.14—0.20)
Head diameter				
Female	—	0.053—0.075	.066 (0.051—0.087)	0.072 (0.068—0.090)
Male	—	0.062—0.075	.054 (0.041—0.071)	0.071 (0.061—0.085)
Head to nerve ring				
Female	—	0.198—0.260	.270 (0.212—0.358)	0.259 (0.231—0.285)
Male	—	0.206—0.224	.222 (0.161—0.294)	0.230 (0.193—0.270)
Length vagina				
% vulva	0.50	—	0.50	0.46 (0.39—0.59)
Egg diameter	0.053—0.068	0.060	0.060	0.053 (0.047—0.065)
Length spicule	0.20	0.20—0.30	0.314 (0.235—0.472)	0.29 (0.20—0.39)
Length male tail	—	0.185—0.264	0.348 (0.170—0.477)	0.354 (0.308—0.470)
Length infective stage	0.25—0.49	0.51—0.58	0.57	0.58 (0.54—0.61)
Width infective stage	0.003—0.012	0.013	0.012	0.013 (0.012—0.015)

TABLE 38

Hosts of *Hydromermis contorta* (Kohn)

Host (Chironomidae)	Stage	Infection[a]	Ref.
Chironomus annularius (Mg.)	L	N	90
C. dorsalis (Mg.)	L	N	90
C. plumosus (L.)	L	N	89
Procladius denticulatus Sub.	L	N	91
Tanytarsus sp.	L	N	91
Zavrelia sp.	L	N	91

[a]   N, naturally occurring infection.

c. Host Range

Hosts of *H. contorta* are listed in Table 38 and consist only of members of the Chironomidae. Attempts to infect larvae of *Aedes aegypti* with this mermithid were unsuccessful.[90] This parasite initiates infection in the host larvae and either emerges from the last larval stage of the midge or is carried into the adult host. In the latter case, host intersexes are formed which are discussed in detail by Wülker[84] and Götz.[82]

d. Culture and Application

Since most of the midges mentioned in the host list can be grown under laboratory conditions, it should be possible to mass produce *H. contorta* to obtain material for field releases. At this time, the present author is not aware of any attempts to use this nematode as a control agent.

D. Mermithid Parasites of Blackflies (Simuliidae)

Due to their painful bite and ability to serve as intermediate hosts of vertebrate parasites, blackflies are medically important insects that warrant control in many areas. Mermithid nematodes are one of the most abundant parasites of blackflies, and for this reason they have attracted attention for the past three decades. A range of mermithid species is known to parasitize at least 64 species of blackflies worldwide.[6]

Mermithids show promise as biocontrol agents of simuliids because of their host specificity and ability to kill or sterilize blackfly larvae or adults, respectively. Extensive studies showed how mermithids have eradicated blackflies from certain areas in the U.S.[92] and the U.S.S.R.[93]

One of the problems that has retarded the practical aspect of using mermithids for blackfly control in comparison with progress made with *Romanomermis culicivorax* and mosquitoes is the difficulty of obtaining enough material for large-scale studies and field releases. Since most blackflies cannot be carried through their entire life cycle under laboratory conditions, in vivo methods of culturing the mermithids are restricted. Unfortunately, in vitro methods of mermithid cultivation are not yet complete; thus, costs and time involved in obtaining mermithids for large-scale field releases are prohibitory at this stage. However, limited field releases have been made in New York with encouraging results. Other laboratories are working hard toward the goal of using native mermithids for the control of blackflies in Canada and for the control of *Simulium damnosum,* the carrier of the parasite causing Onchocerciasis, in West Africa.

1. Mesomermis flumenalis Welch 1962 (syn. Neomesomermis flumenalis (Welch))

The new genus *Neomesomermis* was erected by Nickle[3] to replace *Mesomermis* because the latter was insufficiently described. However, although Daday[94] did not designate a type species when he erected the genus, a type *M. zschokkei* was later designated by Chitwood.[95] Welch[96] gave a complete redescription of *Mesomermis* and it seems unnecessary to add another generic name to replace one that is well defined. For this reason, the genus *Neomesomermis* Nickle is regarded as a synonym of *Mesomermis. Mesomermis flumenalis* was described from larvae of *Simulium venustum* from Ontario, Canada but appears to be distributed throughout North America.

a. Description

The genus *Mesomermis* Daday is characterized by the possession of six hypodermal cords; a straight, barrel-shaped vagina; paired, parallel, separate spicules of medium length, and six cephalic papillae. There are no cross fibers in the cuticle. The following description of *M. flumenalis* is based on the description by Welch[96] and Ebsary and Bennett.[97]

Adults — The mouth is terminal and the pharyngeal tube extends less than one third of total body length. The tail is cone-shaped to rounded. The vestigial excretory pore opening is 280 to 350 μm from the head. Amphids are located slightly dorsal and behind lateral papillae. In the female, the amphidial pouch is circular, with a round pore. The vagina is short and barrel-shaped, lying at right angles to the body axis. In the male, the amphidial pouch is ellipsoid, with elongate elliptical pore. Spicules are separate, parallel, and about twice as long as the anal body width; tips of the spicules are rounded and bordered by small spines. Qualitative characters for the adults are given in Tables 39 and 40.

Preparasitic juveniles — This stage is 1.04 to 1.28 mm long, and the greatest width is 7.5 to 10.3 μm. Stylet length is 3.2 to 4.5 μm; tail length is 5.4 to 10.6 μm.

Postparasitic juveniles — The tail is cone-tipped (sometimes with a long cuticular filament).

TABLE 39

Measurements of Female *Mesomermis flumenalis* [96,97]

Character	Value
Length (mm)	10.0—18.0
Greatest width	130—292
Length amphidial pouch	12—23 × 7—17
Diameter of amphidial opening	7—10 × 5—13
Head to nerve ring	180—210
Length vagina	80—100
% vulva	47—55
Egg diameter	85—94 × 75—81

TABLE 40

Measurements of Male *Mesomermis flumenalis* [96,97]

Character	Value
Length (mm)	6.7—12.0
Greatest width	65—155
Length amphidial pouch	26—32 × 17—27
Diameter of amphidial opening	21—32 × 6—17
Head to nerve ring	180—210
Length spicules	150—242
Width spicules	10—12
Length tail	205—327
Width tail	80—142

Eggs — The shell is 1 μm thick. The eggs are unornamented, covered with a sticky substance attaching them to the substrate, and generally unembryonated when laid.

b. Bionomics

Life cycle[97,98] — At 18° C, the postparasites of *M. flumenalis* molted to the adult stage in 10 to 13 days and began mating within the next few hours. Mating was usually completed in 12 to 24 hours, and oviposition was completed 36 to 75 days after mating. From 600 to 650 eggs were deposited by each of two females. Eggs maintained at 12°C hatched in 35 to 55 days, but the preparasitic juveniles died 2 to 3 days later in the absence of a host. The optimum developmental temperature for this nematode was 12°C, which corresponds to the temperature in the streams where *flumenalis* occurs. In nature, *M. flumenalis* is considered a univoltine species which infects blackflies only from February to July at the latitude of Newfoundland. Highest rates of infection of *Prosimulium* larvae occurred in early May and of *Simulium* populations in late June. This suggests that there are two periods of egg hatching, one in late fall (infecting *Prosimulium* spp.) and one in early spring (infecting *Simulium* spp.).

M. flumenalis generally completes its development and emerges from the host before the latter has pupated; however, parasitism of adult hosts has been observed.[99] Distribution of the mermithid in the stream may arise from movement of parasitized larvae as well as from the flight of infected adult blackflies.

The preparasitic stages of *M. flumenalis* enter larvae of *S. venustum* by directly penetrating through their cuticle.[100] Molloy and Jamnback[101] confirmed this finding by showing that the preparasites directly penetrated through the cuticle of the *S. vittatum* larva to enter the hemocoel. In laboratory experiments, the latter authors obtained 80% infection in first instar *S. vittatum,* 64% for second instar larvae, and 0% for the pupae. The authors emphasized the importance of detritus in the habitat as a substrate on which the preparasites could crawl without being carried downstream. Thus, the long awaited question of how mermithids entered blackfly larvae in fast flowing water was now answered. Direct penetration was the normal method, at least for *M. flumenalis.*

The effect of parasite burden on the sex ratio of *M. flumenalis* was demonstrated by Ezenwa and Carter.[102] Larvae of *P. mixtum/fuscum* and *S. venustum* with one, two, three, and four nematodes per host produced approximately 12, 44, 82, and 94% male mermithids, respectively.

Physiological studies on the effect of *M. flumenalis* on host simuliid larvae show a significant increase in volume of the corpus cardiacum cells of *S. venustum.* This was accompanied by a significant increase in nuclear DNA/RNA activity in the corpus allatum. A lack of any changes in the endocrinology of *P. mixtum/fuscum* indicated that the nematode was not actively modifying the hormonal balance of the host.[103] The latter authors noted that *M. flumenalis* drastically reduced the net amount of body fat in the host larvae, resulting in a significant reduction in fat body glycogen reserves. They concluded that nematode depletion of hemolymph and body fat reserves removed precursors that are required for the synthesis of cuticular proteins during molting.

Natural occurrence[104-108] — In regards to the mermithid parasites of blackflies, more information is available on the natural occurrence of *M. flumenalis* than on any other mermithid nematode. This is mainly due to the studies made on this nematode in Newfoundland by several authors.[104-108] In Newfoundland, *M. flumenalis* is considered a univoltine species that primarily parasitizes the larval stages of *P. fuscum/mixtum* in the winter and *S. venustum* in the early spring. The nematode is widespread, and because of its abundance is considered the most important mermithid parasite of Newfoundland mermithids. It was found in 56 streams out of a total of 59 sampled which contained mermithid parasites and in a total of 198 streams which contained blackflies. It is well adapted to low temperatures, and infection occurs at 4 to 12°C.

Although variables affecting host infections (different breeding sites, location in breeding sites, host developmental stages) made sampling difficult, the average incidence of infection for *M. flumenalis* usually varied from 1 to 11%, but a 20% infection rate was recorded for some blackfly populations. Data concerning infection of adult blackflies is not clear, although *M. flumenalis* has been recovered from *S. verecundum* adults, and infection of this stage may be fairly common.

Near Churchill Falls, Labrador, 5 of 18 streams surveyed contained *M. flumenalis* infections. Out of 3144 blackfly larvae examined, 11.2% were infected with this mermithid. It would appear that *M. flumenalis* is well adapted to blackfly populations in extreme northern latitudes.

c. Host Range

The host range of *M. flumenalis* is shown in Table 41. Only members of the Simuliidae have been found serving as hosts for this species. Although primarily a larval parasite, this nematode has been found in the adult stage of at least one blackfly species and probably occurs in adults of most or all of its range of hosts. It is interesting to note that this species has been reported parasitizing blackflies in the U.S.S.R. as well as in North America.

d. Culture

Laboratory culture of mermithid parasites of blackflies is still in its infancy because of the difficulty in keeping the various host stages alive throughout their development. Even collecting viable nematodes from field-infected blackflies presents difficulties in

TABLE 41

Hosts of *Mesomermis flumenalis*

Host	Stage	Infection[a]	Ref.
Prosimulium mixtum/fuscum Syme and Davies	L	N	92
Simulium corbis Turinn	L	N	109
S. cryophilum (Rubt.)	L	N	93
S. decorum Walker	L	N	109
S. nolleri Fried	L	N	93
S. latipes (Mg.)	L	N	110
S. tuberosum (Lundstrom)	L	N	104
S. venustum Say	L	N	96
S. verecundum Stone and Jamn.	L, P, A	N	108
S. vittatum Zett.	L	E	101

[a] N, natural infection; E, experimentally induced infection.

the maintenance of host larvae for emergence of the postparasitic mermithids.[111] Field samples of larvae should be kept cool and aerated while being transported back to the laboratory. Once in the laboratory, the parasitized blackflies can be placed in containers of aerated water held at 10°C for nematode emergence. The above authors collected 4378 mermithids from approximately 250,000 blackfly larvae amassed over a 4-month period from Newfoundland streams.

A laboratory method of rearing *M. flumenalis* from postparasitic juveniles through the adult to the preparasitic juveniles was discussed by Bailey et al.[111] These authors placed 50 pairs of nematodes in containers with about 450 cc of wet sand. Molting and mating followed with about 79% of the females eventually reproducing. Differences in time of development between nematodes reared from *Simulium* and *Prosimulium* hosts suggests that two physiological strains of *M. flumenalis* may exist.

e. Field Trials

Preparasites of *M. flumenalis* were released against blackfly larvae in a small stream in Cambridge, New York.[112] At the five sampling points, the stream averaged 21 cm (15 to 25) in width, 5.5 cm (3.5 to 8.0) in depth, and 33.4 cm/sec (25 to 41) in current velocity. Preceding the experiment, the stream was heavily stocked with *S. venustum* and *S. vittatum* eggs. Over a period of 4 consecutive days, approximately 1.5 million *M. flumenalis* preparasites were released into the stream. Early instar *S. venustum* which were collected immediately below the treatment point showed an infection rate of 71.4% (N = 49). The infection was higher in *S. venustum* than in *S. vittatum*, indicating a host preference. The authors calculated that to achieve a parasitism rate of 50%, approximately seven preparasites per square centimeter of substrate were necessary. In order to obtain a 90% parasitism rate, about 3.6×10^9 preparasites per mile of stream would be necessary. Cost of the present rate of production ($300.00 per million preparasites) would prohibit the use of *M. flumenalis* until a cheaper method of production could be established.

2. Gastromermis Micoletzky 1923
a. Description

Characters of the genus *Gastromermis* are an S-shaped vagina, six head papillae, a single spicule more than twice the anal body width, six or eight hypodermal cords, mouth usually shifted ventrally, and the adult cuticle lacking apparent cross fibers.

3. Gastromermis viridis Welch 1962[96,92]

This species was described from the blackfly, *S. vittatum,* in Wisconsin and is the

TABLE 42

Measurements of Female *Gastromermis viridis* [96]

Character (N = 12)	Value
Length (mm)	17.3 (10.0—46.1)
Greatest width	250 (170—250)
Head to amphidial opening	19
Length amphidial pouch	12 × 21
Diameter of amphidial opening	8
Head to nerve ring	242 (200—250)
Length vagina	450 (440—580)
% vulva	48 (43—56)
Egg diameter	40—50 × 50—70

TABLE 43

Measurements of Male *Gastromermis viridis* [96]

Character (N = 3)	Value
Length (mm)	11.5 (11.5—16.9)
Greatest width	130—160
Length amphidial pouch	28
Diameter of amphidial opening	12
Head to nerve ring	230—290
Length spicules	800—1000
Widtl. spicules	15—21
Length tail	190—240
Width tail	150 (150—180)

only described species in this genus attacking blackflies in North America. Other species attack European and African blackflies.[1]

a. Description

Adults — Live specimens are greenish in color. The mouth is slightly displaced ventrally and not positioned posterior to the cephalic papillae. Amphids are ovoid and located slightly posterior to head papillae. The excretory pore is 250 to 300 μm from mouth. The pharyngeal tube extends past "midbody" and the tail is rounded. In the females, the amphidial pore is circular and vulvar lips are present. In the male, there is a long, single spicule. Quantitative measurements of the adults are presented in Tables 42 and 43.

Preparasitic juveniles — This stage is 426 (398 to 453) μm long and 12 (8 to 14) μm wide.

Postparasitic juveniles — The width of the tail is appendage 70 to 160 μm long; amphids are smaller than those of the adult.

Diagnosis — The long spicule, slight ventrally displaced mouth, and long tail appendage are characteristic characters of *G. viridis,* as well as the green color.

b. Bionomics

Life cycle[92] — The final molts of *G. viridis* occurred 5 to 12 and 12 to 18 days after emergence from the host of 23.9 and 15.6°C, respectively, and mating occurred immediately afterwards. Eggs were produced 2 to 6 and 13 days after the final molt at 23.9 and 15.6°C, respectively. The eggs were covered with a gelatinous substance which attached them to the substrate. Movement of the embryo was first observed 6 days

TABLE 44

Host Range of *Gastromermis viridis*

Host	Stage	Infection[a]	Ref.
Prosimulium mixtum/fuscum Syme and Davies	L	N	107
Simulium corbis Twinn	L	N	107
S. venustum Say	L	N	107
S. vittatum Zett.	L, P, A	N	92

[a] N, natural infection.

after oviposition, and after hatching the preparasitic juveniles lived 1 to 2 and 3 days at 23.9 and 15.6°C, respectively. They swam about in a haphazard fashion after hatching but became sluggish on the second day, and it was doubtful whether they could enter a host. The manner of entry was presumed to be via the intestine.

Parasitic development took 10 to 14 days at 23 to 27°C, but it took up to 5 months in overwintering blackfly larvae. The stylet became lost during parasitic development, and a caudal appendage formed. During the final molt of the postparasitic juveniles, the cuticle ruptures about one third of the body length posterior to the mouth. Apparently two cuticles are shed at this time.

Natural occurrence[92] — *G. viridis* could be found in parasitized blackfly larvae in all months except April and may typically survive over winter as a parasitic juvenile. This stage has been recovered under rocks and in algal mats growing on the rocks in the stream. Superparasitism was common, and there is a tendency for more males to be formed under these circumstances; however, a 1:2 male to female ratio seemed to persist in some streams.

Rates of parasitism of *G. viridis* are difficult to determine from the data presented by the above authors since their tables include both the above species and *Isomermis wisconsinensis* in their counts. The overall infection rate of *Simulium vittatum* larvae by both mermithids varied from 0 to 95%, depending on the location and time of sampling. In many streams, the first generation of blackflies was heavily parasitized but the later generations less so, indicating that nematode reproduction lagged behind that of the blackflies. Overall larval mortality from nematodes was judged to be about 50%. *G. viridis* also was collected from the pupae and adults of *S. vittatum,* with infection rates in the latter stages varying between 37 and 63%, respectively. Aside from killing the adult flies during emergence, as with host larvae, the nematodes also sterilize the females. In one instance, virtual elimination of blackfly populations was recorded in Wisconsin, primarily as a result of parasitism by *G. viridis.* After high rates of parasitism, no additional host larvae were found on vegetation in this section of the creek.

Aside from Wisconsin, *G. viridis* has also been recovered from Newfoundland, where it has a single generation each year.[106] Its distribution in Newfoundland is sparse if the results of Ebsary and Bennett[104] are indicative of its abundance in other streams. The above authors found this nematode in only 2 out of 59 streams which contained mermithid-infected simuliids. It has also been collected in Ontario.[96]

c. Host Range

The known host range of *G. viridis* is rather limited (Table 44). It appears that *S. vittatum* is the preferred host and the parasite occurs in the larvae, pupae, and adult flies.

d. Culture and Possible Application

Phelps and DeFoliart[92] succeeded in carrying *G. viridis* through at least one complete

TABLE 45

Measurements of Female *Isomermis wisconsinensis* [96]

Character (N = 5)	Value
Length (mm)	14.7 (12.2—21.0)
Greatest width	160 (130—230)
Length amphidial pouch	14
Diameter of amphidial opening	5—7
Head to nerve ring	140—250
Length vagina	240—280
% vulva	49 (42—52)
Egg diameter	37—39×50—57

TABLE 46

Measurements of Male *Isomermis wisconsinensis* [96]

Character (N = 5)	Value
Length (mm)	11.5 (9.1—13.3)
Greatest width	140 (120—140)
Length amphidial pouch	19
Diameter of amphidial opening	3—5
Head to nerve ring	190—250
Length spicules	180 (140—180)
Width spicules	13 (11—15)
Length tail	320 (310—330)
Width tail	140 (110—160)

generation in the laboratory, and therefore this nematode could be cultured under in vivo conditions, although its usefulness may be restricted against just one or two species of blackflies. Additional studies are warranted to determine why other species of blackflies in infested streams are not parasitized.

4. Isomermis Coman 1953
a. Description[96,92]
The genus *Isomermis* is characterized by an S-shaped vagina; paired, parallel spicules between 1 and 2.5 times anal body width in length; eight hypodermal cords; adult cuticle lacking apparent cross fibers, and six cephalic papillae.

5. Isomermis wisconsinensis Welch 1962
This species is known to attack blackflies in North America and was described from *Simulium vittatum* in Wisconsin.[96]

a. Description
Adults — The head is set off by neck constriction. Amphids are ovoid, slightly dorsolateral in position, and situated in the neck region. Cervical papillae open laterally, dorsally, and ventrally. The excretory pore is 220 to 250 μm from head and the mouth is terminal. The pharyngeal tube extends past midbody. The tail is coneshaped and tapered to rounded at the tip. Quantitative measurements of the adults are presented in Tables 45 and 46.

Preparasitic juveniles — This stage is 268 (254 to 277) μm long and 16 (14 to 21) μm wide.

Postparasitic juveniles — This stage has a tail with a caudal appendage, 10 to 18 μm long.

TABLE 47

Host Range of *Isomermis wisconsiensis*

Host	Stage	Infection[a]	Ref.
Prosimulium mixtum/fuscum Syme and Davies	L	N	107
Simulium venustum Say	L	N	104
S. vittatum Zett.	L, P, A	N	92

[a] N, natural infection.

b. Bionomics

Life cycle[92] — The final molts of *I. wisconsinensis* occurred 2 to 5 and 6 to 9 days after emergence from the host at 23.9 and 15.6°C, respectively, and mating occurred at once. Eggs were produced 2 to 5 and 2 to 8 days after the final molt at 23.9 and 15.6°C, respectively, and egg laying continued for 3 to 4 days. One female deposited 1200 eggs; the eggs were sticky and adhered to the substrate. The first movements of the embryo were noted 6 days after oviposition.

The preparasitic juveniles survived 1 to 2 and 3 days at 23.9 and 15.6°C, respectively, although they became sluggish on the second day and may not have been infective. It was presumed that they entered the host via the intestine. Parasitic development took 10 to 14 days at 23 to 27°C but up to 5 months in overwintering blackfly larvae. The final molt of the postparasitic juveniles appeared to be double with the cuticle rupturing about one third of the body length posterior to the mouth.

Natural occurrence[92] — *I. wisconsinensis* could be found parasitizing blackfly larvae only from June to November and probably survives over winters as a postparasitic juvenile. The postparasitic juveniles have been recovered under rocks and intertwined with algae growing on rocks in the stream. Super parasitism was observed with the tendency for more males to be produced as the parasite burden increased. However, a male to female ratio of 1:2 was observed in some streams.

Parasitism rates of *I. wisconsinensis* are difficult to determine from the data presented by the above authors since their tables include both the above species and *Gastromermis viridis* in their counts. However, the overall infection rate of *Simulium vittatum* larvae by both mermithids varied from 0 to 95%, depending on location and time of sampling. The first generation of blackfly larvae was heavily parasitized, but the incidence of infection declined, suggesting that the blackflies reproduced faster than the nematodes. Overall, larval mortality from nematodes was judged to be 50%. *I. wisconsinensis* also was recovered from pupae and adults of *S. vittatum*, with infection rates in the latter stages between 37 and 63%. Parasitized female flies are sterilized by the mermithid and are killed upon emergence of the host, as is typical for larval blackflies. In one instance, virtual elimination of *S. vittatum* larvae resulted from high rates of nematode parasitism, especially by *I. wisconsinensis*.

Aside from Wisconsin, this mermithid has also been recorded from Newfoundland, where it is considered to have a single cycle each year.[106] Apparently, its distribution in Newfoundland is sparse, since Ebsary and Bennett[104] recovered *I. wisconsinensis* in only 4 out of 59 streams containing mermithid infected simuliids.

c. Host Range

The known host range of *I. wisconsinensis* is quite limited (Table 47), and it appears that *S. vittatum* is the preferred host. The nematode occurs in the larval, pupal, and adult stage of this blackfly.

TABLE 48

Measurements of *Isomermis lairdi* Females[113]

Character (N = 12)	Value
Length (mm)	14.5 (8.7—18.6)
Greatest width	191
Head to amphidial opening	22
Length to amphidial pouch	12
Diameter of amphidial opening	5—6
Head to nerve ring	212
Length vagina	283 (200—343)
% vulva	51 (48—57)
Egg diameter	60 × 65

TABLE 49

Measurements of *Isomermis lairdi* Males[113]

Character (N = 20)	Value
Length (mm)	9.9 (5.2—14.4)
Greatest width	128
Head to amphidial opening	16
Length amphidial pouch	15
Diameter of amphidial opening	10
Length spicules	219 (187—285)
Width spicules	7
Length tail	226
Width of tail	122

d. Culture and Possible Application

Since Phelps and DeFoliart[92] succeeded in laboratory infections and parasitic development of *I. wisconsinensis,* this nematode could be cultured under in vivo conditions. Its usefulness may be restricted to just one or two blackfly species because of a possible limited host range. Further studies are needed to elucidate the defense mechanisms of other blackfly species to this nematode.

6. Isomermis lairdi Mondet, Poinar, and Bernadou 1977

This parasite is one of the most common mermithids found attacking all stages of *Simulium damnosum* in portions of West Africa.

a. Description[113]

Adults — The mouth is terminal and amphids are situated just behind the row of cephalic papillae, with amphidial openings round, and shifted slightly dorsad. In the females, the excretory pore is located 3.9% (3.0 to 5.1) of the body length from the head end. The vagina is strongly S-shaped and vulva lips are absent. In the male, the excretory pore is located 4.2% (3.3 to 5.0) of the body length from the head end. Genital papillae are arranged in three rows, the middle row bifurcating around the cloacal opening. Quantitative values of adults are presented in Tables 48 and 49.

Preparasitic juveniles — This stage is 500 μm long and 15 μm wide. Stylet, paired penetration glands, stichosome, intestine, and anus are conspicuous. The tail is relatively short.

Postparasitic juveniles — This stage has a tail appendage 10 to 15 μm long.

b. Bionomics[114,115]

Life cycle (at 23 to 28°C) — Postparasitic juveniles molt (double molt) to the adult stage 3 to 4 days after leaving their host. Although found in warm tropical waters, the adults of *I. lairdi* could remain alive for 8 days at 4°C. Mating occurs immediately afterwards, and oviposition begins 10 to 12 days after the postparasitic juveniles emerge. One female layed 500 eggs. The eggs are covered with a sticky substance which attaches them to the substrate. After molting once in the egg, the preparasitic juveniles hatched 5 to 8 days after oviposition.

They were very active and entered simuliid larvae by direct penetration, primarily around the mouthparts of the host, never from the alimentary tract. Parasitic development lasted from 10 to 16 days, with the entire life cycle occurring between 25 and 39 days. When the parasite is carried into the adult stage of the host, it leaves about 5 to 7 days after the host takes a blood meal. Parasites leaving adult hosts show further development of their gonads than those leaving larval blackflies (Figures 18 and 19).

Natural occurrence — *I. lairdi* is one of the most commonly encountered parasites of blackflies in the Ivory Coast, West Africa. In the field, *I. lairdi* is estimated to have six to eight biological cycles each year. Postparasitic juveniles and adults at various levels of development can be obtained from moist sand in the river beds during the West African dry season.

Rates of parasitism in adult *S. damnosum* reached 80% of the nulliparous females and 35% of the entire population after mermithid populations built up. In such cases, atrophy of the ovaries occurred in 98.9% of the individuals. While working in West Africa on this problem, the present author noted that all records indicated that only nulliparous adult flies were infected by *I. lairdi* (as determined by the reduction of body fat and elasticity of ovaries). Since parasitized flies have reduced gonads and fat body, they probably don't deposit eggs; however, they still can take at least one blood meal, which may be necessary for the mermithids to mature inside the adult flies. Whether a single blood meal is sufficient for the maturation of most mermithids may depend on parasite burden and age. It was also significant that when two or more mermithids were present in adult flies, one was usually a female and the others males. Aside from the Ivory Coast, *I. lairdi* also was recovered from female *S. damnosum* in northern Ghana.[115a]

c. Host Range

Aside from *S. damnosum*, *I. lairdi* also has been found parasitizing other blackflies in the same habitat (Table 50). Only the adults of *S. damnosum* have been found infected, but there is no apparent reason why this nematode should not occur in the adults of other blackfly hosts. Aside from the laboratory infections described above, *I. lairdi* has not yet been cultured extensively for field releases.

7. Romanomermis culicivorax Ross and Smith 1976

Although this nematode develops only in mosquito larvae, it will attack and penetrate insect representatives in other families of Diptera, including the Simuliidae. Thus, Finney[40] showed that the preparasites of *R. culicivorax* would enter the body cavity of first, second, and third instar larvae of *Simulium venustum, S. decorum,* and *S. vittatum.* It was later shown that *S. damnosum* was similarly infected, even in moving water systems in the laboratory[29] (Figure 20). Although the latter authors noted some initial growth of the parasite in laboratory infected *S. damnosum,* the hosts always died before there was an opportunity to determine if nematode development could be completed. However, by using a special retaining chamber placed in a moving water system, Hansen and Hansen[115b] were able to keep the hosts alive for a longer period than with other methods. The acrylate plastic chamber was 7 mm in diameter and 22

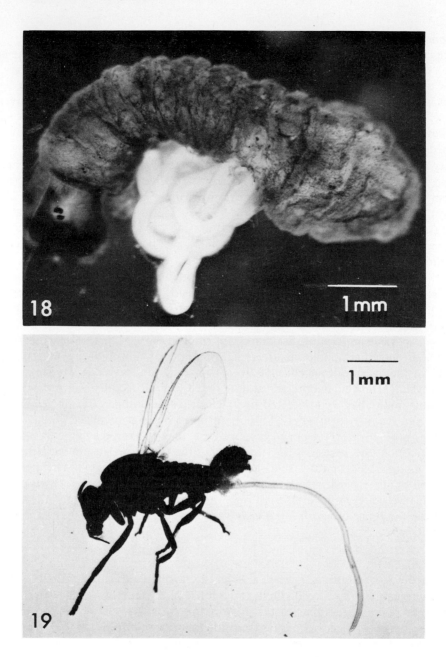

FIGURE 18 and 19.   (18) A larva of *Simulium damnosum* parasitized by *Isomermis lairdi*. (19) Adult *Simulium damnosum* parasitized by *Isomermis lairdi*.

mm long; it was closed at both ends with removable polyester screens (60 to 400 μm mesh — depending on the size of larvae used). Aquarium water was drawn through the chamber by an air lift pump at about 50 cm/sec velocity. First stage larvae of *S. damnosum* attached to the walls or screen of the chamber and reached pupation 9 days after hatching. While in the chamber, the simuliids could be exposed to the preparasitic stages of mermithid nematodes (Figure 21).

Although the development of *R. culicivorax* was never completed in any of the blackflies infected and it is questionable that normal development could take place in members of this family, the nematode still has a decided effect on the host.

TABLE 50

Simuliid Hosts of *Isomermis lairdi* [114]

Host	Stage	Infection[a]
Simulium adersi Pomeroy	L	N
S. alcocki Pomeroy	L	N
S. cervicornutum Pomeroy	L	N
S. damnosum Theobald	L, P, A	N
S. unicornutum Pomeroy	L	N
S. vorax Pomeroy	L	N

[a] N, natural infection.

 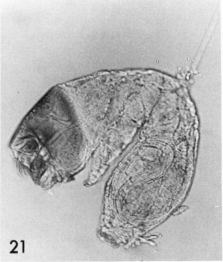

FIGURE 20 and 21.   (20) Plastic chamber and tubing used for infecting *Simulium* larvae. (Courtesy of E. and J. Hansen.) (21) Preparasitic juvenile of *Romanomermis culicivorax* entering a first stage *Simulium damnosum* larva. (Courtesy of E. Hansen.)

In regards to *S. damnosum*, Hansen and Hansen[115b] found that every larva that was penetrated by a preparasite of *R. culicivorax* died between 20 min and several hours after the attack. Only the first and second instar host larvae were readily attacked, and in field trials (see section on *R. culicivorax*) they calculated that for every host penetrated, another two or three were killed by the preparasites simply damaging the cuticle without entry. In the latter cases, the hosts would detach, float downstream, and eventually perish.

The description of this nematode and other facts about its host and field application are given in the section on mosquito parasites.

E. Terrestrial Mermithids

Mermithids parasitize a wide range of terrestrial insects, although the biology of the nematode has been studied in relatively few cases. In general, published reports constitute a description of the parasite followed by the stage of host found infected and, at most, a few observations on the incidence of infection and method of parasite exit. Relatively few terrestrial mermithids have been carried through their entire life cycle

TABLE 51		TABLE 52	
Measurements of *Mermis nigrescens* Females[121,124-126]		Measurements of *Mermis nigrescens* Males[121,124-126]	
Character	Value	Character	Value
Length (mm)	110 (67—163)	Length (mm)	40—58
Greatest width	466	Greatest width	200
Head to nerve ring	428	Head to nerve ring	240
Length vagina	404	Length spicules	200
% vulva	48 (41—50)	Length tail	200
Egg diameter	38—54 × 48—56	Width tail	200

in the laboratory. This is partly due to the difficulty of providing conditions necessary for infection, something that usually presents no serious problems with mermithids of aquatic hosts. However, unless infection can be readily achieved under laboratory conditions, there is little hope for mass culture and large scale releases unless in vitro development is possible. The terrestrial mermithids discussed below can be reared under laboratory conditions and have been, if not used on a small scale, then at least considered as practical biological control agents.

1. Diagnosis of Mermis Dujardin 1842[116]

The head has four cephalic papillae, two lateral lip papillae, and two small amphids. The mouth is positioned terminal or slightly ventral. There are six hypodermal cords. The cuticle has distinct cross fibers. The vagina is S-shaped. Eggs may or may not have polar caps or appendages (byssi). Spicules are paired, straight, separate, and approximately equal to anal body diameter. The postparasitic juvenile is usually lacking a tail appendage.

2. Mermis nigrescens Dujardin 1842 (syn. Mermis subnigrescens Cobb 1926, syn. Mermis meissneri Cobb 1926)

This nematode has a unique life cycle that is not known to be paralleled by any other member in the family. The habit of females depositing their eggs on the aboveground portion of plants is an astonishing modification of behavior for insect parasitism. Other species in the genus do not possess these habits.[116]

a. Description[125]

Adults — Characters are as described in generic diagnosis (above). In the female, the vestigial anus is located 870 μm from tail terminus. Two areas in the neck region contain orange red pigment. Eggs are brown with copious branched, dichotomous polar byssi. The embryo is fully formed at time of oviposition. The vagina is S-shaped. In the male, orange-red pigment is almost totally absent. The male contains sperms, 25 μm long; Genital papillae are in three rows, and the middle row bifurcates around the cloacal opening. Quantitative measurements of the adults are given in Tables 51 and 52.

Preparasitic juveniles — This stage is 240 to 340 μm long and 14 to 20 μm wide. The body is relatively short and stout, with a bent stylet approximately 10 to 20 μm long. Paired penetration glands, a schistosome with only eight stichocytes, intestine, anus, and genital primerdium are also present.

Postparasitic juveniles — This stage lacks a tail appendage.

Eggs — With two shells, the outer, dark brown shell contains a fracture line across the middle and two opposite poles bear an extension of copious branched, dichotomous byssi (serving to attach the eggs to a leaf substrate). The inner shell is light brown and membranous (Figure 22).

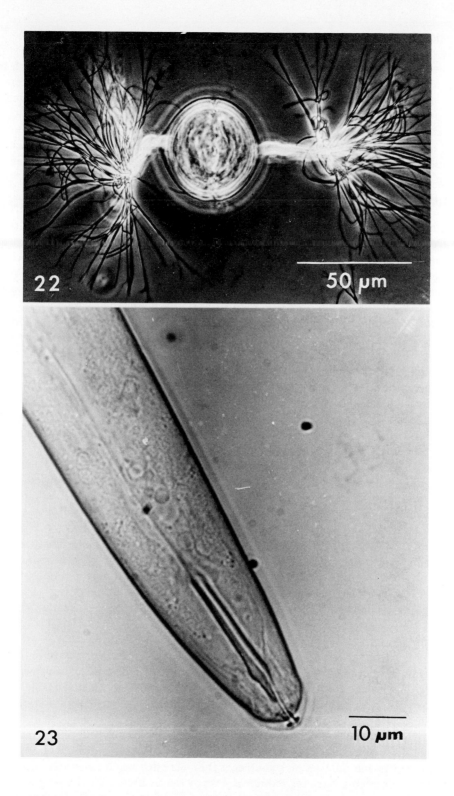

FIGURE 22 and 23.   (22) Egg of *Mermis nigrescens* showing polar appendages. (23) Head of a molting preparasitic juvenile of *Filipjevimermis leipsandra* just after hatching.

b. Bionomics[118]

Life cycle — Females of *M. nigrescens* ascend vegetation during or just after spring showers to deposit their eggs. A single female contains up to 14,000 eggs which may be deposited at the rate of 1000 per minute. Shortly after the sun emerges, the females return to the soil, although multiple ovipositions are possible. The eggs, which are fully embryonated at deposition, are laid on vegetation which is later eaten by a grasshopper host. Once inside the host's intestine, the second stage juvenile (one molt occurs in the egg) hatches and penetrates through the gut wall into the hemocoel. The parasitic male nematodes usually develop in 4 to 6 weeks, whereas the females may take up to 10 or 12 weeks depending on temperature and host size.

The postparasitic juveniles then leave the host, which soon dies, and enter the soil where a long period ensues before the final molts to the adult stage. This occurs 2 to 4 months later; then the females remain in the soil for another 4 to 6 months before they are sexually mature. Mating may occur, but males are uncommon and not necessary for egg production since most females seem to reproduce parthenogenetically. Egg production may continue for a year, and females may live up to 3 years. This is a very long survival period for members of this family. The adults are found at various depths in the soil but the majority occur from 6 to 18 in. below the surface.

Natural occurrence — *M. nigrescens* is common in the British Isles, northern Europe, and North America. The incidence of infection is difficult to determine since most reports simply indicate grasshopper parasitism by mermithids, and there are other mermithids, aside from *M. nigrescens*, that attack grasshoppers.

Christie[118] found an average mermithid infection of 12% in 2500 grasshoppers examined from New Hampshire, and the majority were considered *M. nigrescens*. Briand and Rivard[119] reported 12 to 23% infection of *M. nigrescens* in grasshoppers from Quebec, and Glaser and Wilcox[120] recorded 12 to 76% parasitism of *Melanoplus bivittatus* by a nematode fitting the description of *Mermis nigrescens*.

In studying the mermithid parasites of grasshoppers in Iowa, Denner[121] concluded that the great majority of infections were due to *M. nigrescens* and recorded a 15 to 18% infection rate in representatives of the families Acrididae and Tettigoniidae. Up to 70% of the females and 52% of the males of the grasshopper *Hesperotettrix viridis pratensis* were infected with *M. nigrescens* in southeastern North Dakota.[122] On the other hand, parasitism of *Melanophus bivittatus* in the same area did not exceed 5% of either male or female populations. However, the authors concluded that one population of *H. viridis pratensis* was completely destroyed prior to egg laying by *Mermis nigrescens* and suggested the use of this nematode to control grasshoppers, especially in a damp habitat, since that is where they noticed the most severe parasitism.

c. Host Range

In nature, *M. nigrescens* is mainly a parasite of grasshoppers; however, there are records of its development in Lepidoptera and other insect orders. Thus, this nematode apparently has a wide host range, perhaps restricted mainly by host size since development can probably only be completed in large-bodied insects whose period of development is several weeks or longer. A list of the hosts is presented in Table 53. It is not too difficult to explain parasitism in those insects which feed on plant leaves that could have been covered with eggs of *M. nigrescens*. In the case of honey bee parasitism, it is possible that the bees picked up eggs that were deposited on flowers. In female grasshoppers, the nematode may reduce or destroy the ovaries, while in male grasshoppers the testes may be reduced or not noticeably affected. Infected grasshoppers die either before or after the nematodes emerge.

d. Culture and Possible Application

There is general agreement that *M. nigrescens* can be an important agent for con-

TABLE 53

Hosts of *Mermis nigrescens*

Order	Family or superfamily	Host	Stage	Infection[a]	Ref.
Orthoptera	Acridoidea	*Arphia sulphurea* (Fab.)	J, A	E	118
		Camnula pellucida (Scudder)	J, A	N	119
		Chorthippus albomarginatus (DeG.)	J, A	N	123
		C. biguttulus (L.)	J, A	N	124
		C. curtipennis (Harris)	J, A	N	125
		C. longicornis (Latreille)	J, A	N	118
		C. longipennis Scudder	J, A	N	125
		C. parallelus (Zett.)	J, A	N	126
		C. scalaris F. W.	J, A	N	123
		C. variabilis Fieb.	J, A	N	123
		C. viridifasciata (DeG.)	J, A	N	125
		Decticus verrucivorus A.-S.	J, A	N	127
		Dissostevia carolina (L.)	J, A	N	121
		Encoptolopus sordidus (Burm.)	J, A	N	125
		Gomphocerus maculatus Thumb.	J, A	N	123
		G. sibiricus (L.)	J, A	N	126
		Hesperotettix viridis pratensis Scudder	J, A	N	122
		Locusta migratoria migratorioides R. and F.	J, A	N	128
		Mecosthetus grossus L.	J, A	N	126
		Melanoplus atlanis (Riley)	J, A	N	118
		M. bivittatus (Say)	J, A	N	118
		M. differentialis (Thomas)	J, A	N	129
		M. femur-rubrum (DeG.)	J, A	N	125
		M. lividus (Dodge)	J, A	N	119
		M. mexicanus (Sauss.)	J, A	N	118
		M. sanguinipes (Fab.)	J, A	N	118
		M. roeseli Hagen.	J, A	N	118
		Orphulella pelidne (Burm.)	J, A	N	125
		O. speciosa (Scudder)	J, A	N	119
		Parapleurus alliaceus (Germ.)	J, A	N	126
		Phoetalictes nebrascensis (Thomas)	J, A	N	121
		Podisma pedestris (L.)	J, A	N	127
		Psophus stridulus (L.)	J, A	N	127
		Romalea microptera (Beauv.)	J, A	E	118
		Schistocerca americana (Drury)	J, A	N	121
		S. gregaria (Forsk.)	J, A	N	128
		Scudderia furcata furcata (Brunner)	J, A	N	121
		Spharagemon collare (Scudder)	J, A	N	119
		Stauroderus scalaris (F.-W.)	J, A	N	130
		Stenobothrus curtipennis Harris	J, A	N	125
	Tettigoniidae	*Chelidoptera roeselii* (Hgb.)	J, A	N	126
		Conocephalus brevipennis (Scud.)	J, A	N	118
		C. fasciatus (DeG.)	J, A	N	119
		C. strictus (Scudder)	J, A	N	121
		Dectius verrucivorus L.	J, A	N	131
		Neoconocephalus ensiger (Harris)	J, A	N	121
		Onconotus laxmanni (Pall.)	J, A	N	132
		Tettigonia viridissima (L.)	J, A	N	131
Dermaptera	Forficulidae	*Chelidurella acanthopygia* (Gene)	—	N	124
		Forficula auricularia L.	—	N	133
Coleoptera	Chrysomelidae	*Agelastica tanaceti* (L.)	—	N	132
	Coccinellidae	*Coccinella septempunctata* L.	—	N	124
	Scarabaeidae	*Melolontha hippocastani* Fab.	—	N	134
		M. melolontha (L.)	—	N	130

TABLE 53 (Continued)

Hosts of *Mermis nigrescens*

Order	Family or superfamily	Host	Stage	Infection[a]	Ref.
Leipdop-tera	Arctiidae	*Arctia caja* (L.)	L	N	135
	Galleriidae	*Galleria mellonella* (L.)	L	E	7
	Geometridae	*Hybernia defoliaria* Clerck	L	N	136
	Lasiocampidae	*Malacosoma castrensis* (L.)	L	N	137
	Lymantriidae	*Liparis chrysorrhoea* (L.)	L	N	126
	Noctuidae	*Plusia chrysitis* (L.)	L	N	127
Hymenop-tera	Apidae	*Apis mellifera* L.	A	N	374

*   N, natural infection; E, experimentally induced infection.

trolling grasshoppers. Laboratory infections are relatively simple. The eggs can be collected by pressing them out of field collected females or brushing them off leaves collected from plants in an infested area. The eggs can be kept viable on a moistened filter paper in the refrigerator for up to one year, and Christie[118] found that eggs kept in a bottle of water in the laboratory remained viable for at least 2 months. These eggs could easily be fed to grasshoppers maintained in the laboratory, and the postparasitic juveniles could be collected. The only disadvantage is the long developmental period of the nematode during both its parasitic and free-living stages. Shipping and applying the eggs to vegetation should present no great difficulties, although if recycling of the parasite is to occur, a region with sufficient rainfall to maintain the free-living stages should be chosen. The present author does not know of any instance where *M. nigrescens* has been used in field trials.

3. *Filipjevimermis leipsandra* Poinar and Welch 1968

This mermithid was discovered attacking *Diabrotica* beetle larvae in a sweet potato field in South Carolina. It has only been reported once in the literature but has many attributes of a good biological control agent.

a. Description[139]

The genus *Filipjevimermis* is characterized by six cephalic papillae, six hypodermal cords, an S-shaped vagina, and an adult cuticle lacking apparent cross fibers. Males, when present have two long, separate spicules. This species is described from females that reproduce parthenogenetically. Only a single male was recovered, and it appeared nonfunctional.

Adults — In the female, the mouth opening is shifted slightly ventrad. Amphids are horseshoe shaped, opening at the level of the cephalic papillae but slightly shifted dorsad. The tail is bluntly rounded, swollen, and possesses reduced bursal muscles. Vestigial anus, male copulatory papillae and excretory pore are present. Quantitative measurements of the females are given in Table 54. The male (N = 1) is 6.3 mm long and the greatest width is 120 μm. There are two separate, long, parallel spicules, one 441 μm and the other 416 μm in length.

Parasitic stages — They develop initially within the host's ganglia, then break out into the hemocoel (see Poinar[140] for an account of the development). **Preparasitic juveniles (Figure 23)** — This stage is 540 (470 to 600) μm long and the greatest width is 18 (17 to 19) μm. The stylet is 20 μm long. A pair of penetration glands, a well-formed stichosome containing 16 stichocytes, an intestine containing a lumen in the anterior portion, an anus, the gonad primordium, and a nerve ring are all present.[142] The intestine is connected to the stichosome in this stage.

TABLE 54

Measurements of *Filipjevimermis leipsandra* Females[139]

Character (N = 15)	Value
Length (mm)	29 (11—47)
Greatest width	157 (87—260)
Length amphidial pouch	12 (10—13)
Head to nerve ring	318 (254—385)
Length vagina	406 (277—501)
% vulva	49 (44—59)
Egg diameter	69 (62—74)

Postparasitic juveniles — This stage has a tail appendage 26 (22 to 28) μm long.

Diagnosis — The shape and location of the amphids, slightly ventral shifted mouth, vestigial anus, and swollen tail separate it from other species in *Filipjevimermis*.

b. Bionomics

Life cycle[140,141] — Females of *F. leipsandra* are able to reproduce parthenogenetically, and males are rarely formed. Each female produced several thousand eggs which hatched in 12 to 15 days at 24°C. After molting once inside the egg, the preparasitic juvenile emerged and actively searched out a host (in this case a larvae of *Diabrotica* sp.).

Entry into the host was achieved by direct penetration, with the assistance of the stylet and glandular secretions. This stage remained infective for 50 days in soil. After entering the host's hemocoel, the nematodes would immediately seek out the ganglia, penetrate through the neural lamella and perineurium, and come to rest in the neuropile. Once inside the ganglion, the parasite initiates development (Figures 24 and 25). Migration into the ganglia protects the nematodes from a lethal defense reaction of the host.

Parasitic development took 12 to 22 days depending on the host instar and number of parasites present. As growth progressed, the neural lamella stretched and finally ruptured, releasing the nematode into the hemocoel. The host died shortly after the postparasitic juveniles made their exit, and the latter molted twice to the adult stage and began depositing eggs immediately afterwards. The parasite is rarely carried over into the adult stage of the host.

Natural occurrence[141] — *F. leipsandra* was originally discovered attacking *Diabrotica* larvae in sweet potato fields in South Carolina. Infection rates of newly hatched *Diabrotica* larvae placed in sweet potato fields ranged from 33 to 100% from August to October. It was noted that increased rates of parasitism were achieved when soil moisture was high.

Parasitism of the flea beetle *Systena blanda* was also noted in field collections, indicating that *F. leipsandra* is not host specific. Field observations suggest that this mermithid is capable of exerting a strong influence on populations of *Diabrotica* spp. and possibly other insects in the habitat. Because of its short generation time (between 1 to 2 months), ability to reproduce parthenogenetically, and ease of rearing under laboratory conditions, *F. leipsandra* represents a promising biological control agent.

c. Host Range

Although *F. leipsandra* was collected from only two genera of chrysomelid beetles in nature (Table 55), laboratory investigations showed that it was capable of developing in other insects as well. Penetration and entry into the ganglia of a midge larva and scarid fly larvae also occurred, although the hosts died before development was completed, and it is not known if these insects could serve as true developmental hosts.

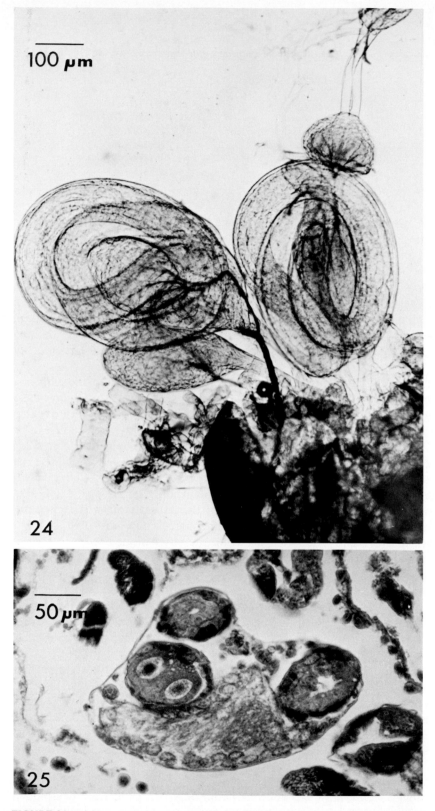

FIGURE 24 and 25.   (24) Parasitic juveniles of *Filipjevimermis leipsandra* developing in the protocerebral lobe and subesophageal ganglion, respectively, of a second instar *Diabrotica* larvae. Dark object is the head capsule of the host. (25) Section through a ganglion of a *Diabrotica* larva showing a developing parasitic juvenile of *Filipjevimermis leipsandra*.

TABLE 55

Host Range of *Filipjevimermis leipsandra*

Order	Family	Host	Stage	Infection[a]	Ref.
Coleoptera	Chrysomelidae	*Diabrotica balteata* Leconte	L	N	139
		D. undecumpunctata howardi Barber	L	E	141
		D. undecumpunctata undecimpunctata	L	E	140
		Systena blanda Melsh	L	N	141
	Curculionidae	*Hypera postica* (Gyll.)	L	E	143
Lepidoptera	Galleriidae	*Galleria mellonella* (L.)	L	E	143
	Noctuidae	*Heliothis zea* Boddie	L	E	143

[a] N, natural infection; E, experimental infection.

d. Culture

Using larvae of *Diabrotica* spp. as hosts, several generations of *F. leipsandra* were established in the laboratory. Infection was obtained by placing 1-day-old host larvae between two layers of moist filter paper in a small petri dish. Preparasitic juvenile nematodes were placed directly on both inner faces of the filter paper. The edges of the filter paper were held together with weights for 80 min, then the insects were removed and kept on roots of germinating corn seedlings. Field releases to control specific insects have not been made with this species, to the present author's knowledge.

4. Agamermis Cobb, Steiner, and Christie 1923
a. Description

The genus *Agamermis* is characterized by six head papillae; an S-shaped vagina; small threadlike amphids; cuticle with cross fibers; two short spicules, which are separate and less than twice the tail diameter, and a caudal scar rather than an appendage on the postparasitic juveniles. The preparasitic juveniles amputate their tail during penetration into the host.

5. Agamermis decaudata Cobb, Steiner, and Christie 1923

This nematode primarily parasitizes grasshoppers in nature, and its development and life history were studied by Christie[144] in one of the most complete investigations ever conducted on an entomogenous nematode. This classical work forms the basis of our understanding about mermithid biology and development. Although this nematode has never been used as a biological control agent in the field, laboratory infections are fairly easy to achieve, and grasshopper hosts can be maintained under laboratory conditions.

a. Description

Adults — The mouth opening is terminal, amphids are threadlike, and the tail is rounded to conical. In the females, the vagina is S-shaped and a cuticularized vulva cone is present. In the male, there are paired, separate, parallel spicules and tail papillae are in four to six broken rows. Quantitative measurements of the adults are shown in Tables 56 and 57.

Preparasitic juveniles — This stage is 5.1 (3.0 to 5.6) mm long and the greatest width is 23 to 25 μm. The stylet is 33 to 37 μm. The posterior portion of its body is amputated during host penetration. Excretory pore, nerve ring, penetration glands, stichosome with 18 stichocytes, genital premordium, and intestine are present.

Postparasitic juveniles — This stage has a caudal tip with a small node rather than a long appendage.

TABLE 56

Measurement of *Agamermis decaudata* Females[117,144]

Character	Value
Length (mm)	30—465
Greatest width	500
% vulva	58 (50—60)
Egg diameter	170 (106—180)

TABLE 57

Measurements of *Agamermis decaudata* Males[117,144]

Character	Value
Length (mm)	10—120
Greatest width	200 (150—250)
Length spicules	175 (150—180)

b. Bionomics[144]

Life cycle — The preparasitic juvenile molts once inside the egg, and after hatching actively locates a grasshopper host. After puncturing the cuticle (takes from 1 to 10 min or longer), juveniles enter the host's body cavity and begin developing in the hemocoel. During entry, the posterior portion of the body is broken and remains outside; only the anterior portion enters and initiates development.

Two types of preparasitic juveniles occur, mainly differentiated by size; however, some differences were also noted in regards to the number of trophocytes, the formation of the node, and the shape of the tail. They probably reflect differences arising from availability of nutrients during egg formation. The length of the parasitic stage is quite variable, but was usually 1 to 1.5 months for males and from 2 to 3 months for females.

Shortly after leaving the host, the postparasitic juveniles enter the soil where they molt, mate, deposit their eggs, and die. The nematodes usually remain in the upper 15 cm of soil coiled together in knots, and molting occurs only 9 to 11 months after the nematodes have left their host. The hosts usually die within 24 hr after the nematodes have left. This is a very long period of maturity for mermithids and can be contrasted with some of the aquatic forms, which require only 1 to 2 days before molting. *A. decaudata* females begin depositing eggs about 1 to 3 months after their final molt. Each female probably deposits about 10,000 eggs. At room temperature, the eggs develop within 50 to 60 days, but the juveniles are able to remain in the egg up to 7 months before hatching, depending on environmental conditions. The amount of nourishment also seems to affect sex determination in this species, e. g., the more parasites per host, the more male nematodes are formed.

Natural occurrence — *A. decaudata* has been recorded only from the eastern U.S.[144] The average rate of infection in over 2000 male and female Acridiidae and Teltigoniidae examined ranged between 10 and 20% (sometimes reaching 25% in some areas), high enough to keep the parasite established, yet low enough to maintain host populations.

A. decaudata, like *Mermis nigrescens,* is long-lived in comparison to many other mermithid nematodes. The postparasites emerge from the host in September, enter the soil, and survive over winter as postparasitic juveniles. Molting occurs in July and the eggs are deposited in August, but they do not hatch until the following spring. Grasshoppers infected in June die when the postparasitic juveniles emerge in September. The same females that deposited eggs in June continue to deposit eggs the following year. Most females do not survive a third winter, and those that do probably die without depositing additional eggs. Males are essential for reproduction and each female is usually accompanied by two to four males, all intertwined in knots in the soil. In the vicinity of Falls Church, Virginia, *A. decaudata* is confined to soil that is usually saturated with water throughout the winter and is frequently flooded for short periods. Although the moisture may retard egg laying, it may be essential for the well-being of the free-living stages.

TABLE 58

Host Range of *Agamermis decaudata*

Order	Family or superfamily	Host	Infection[a]	Ref.
Coleoptera	Coccinellidae	*Ceratomegilla fuscilabris* Mulsant	N	144
Orthoptera	Acridoidea	*Camnula pellucida* (Scudder)	N	144
		Dissosteira carolina (L.)	N	145
		Melanoplus bivittatus (Say)	N	144
		M. femoratus (Burm.)	N	144
		M. femur-rubrum (DeG.)	N	144
		M. mexicanus (Sauss.)	N	144
		M. sanguinipes (Fab.)	N	146
	Tettigoniidae	*Conocephalus brevipennis* (Scud.)	N	144
		Orchelimum vulgare Harris	N	117

[a]   N, natural infection.

c. Host Range

Grasshoppers of the Acridoidea and Tettingoniidae are the preferred hosts of *A. decaudata* (Table 58). Thus, it was surprising that Christie[144] found the coccinellid beetle *Ceratomegilla fuscilabris* infected with this mermithid. It is not known if development could be successfully completed in this beetle, however.

d. Culture

Christie[144] mentioned that the experimental infection of insects with *A. decaudata* was easily accomplished in the laboratory, using a simple type of infection chamber. Since grasshopper hosts are easily maintained under laboratory conditions, it would be possible to rear *A. decaudata* for field releases, although to the present author's knowledge this has never been done.

II. Tetradonematidae

Members of this family are related to the Mermithidae but are not as narrow and elongated as the latter, lack cephalic papillae, and juveniles and adults both occur in the living host. Like mermithids, tetradonematids usually kill their host at the time of emergence and, therefore, are potential biological control agents.

Of the five genera and species in this family, only *Tetradonema plicans* has been used as a biological control agent, although probably all of the species could be cultured under laboratory conditions without much difficulty. *T. plicans* was first found in 1915 parasitizing *Bradysia coprophila* in Kansas.[147] It is apparently restricted to members of the Sciaridae and Mycetophilidae and probably has a world wide distribution.

A. *Tetradonema plicans* Cobb 1919

1. Description

The genus *Tetradonema* (containing only the single species *T. plicans*) is characterized by four large cells (the tetrad) in the pharyngeal region of the female, an acutely pointed tail in the female, and the male tail lacking genital papillae. Descriptions of *T. plicans* have been presented by Cobb,[147] Hungerford,[148] Ferris and Ferris,[149] and Hudson.[150] The following description is taken from the above authors.

Adults — The cuticle is finely striated. Cephalic papillae and amphids are absent, but lateral fields are present. The alimentary tract is degenerate. The pharynx is com-

TABLE 59		TABLE 60	
Measurements of *Tetradonema plicans* Females[147]		Measurements of *Tetradonema plicans* Males[147]	
Character	Value	Character	Value
Length (mm)	0.870—14.000	Length (mm)	0.434—0.800
Greatest width	120	Greatest width	42
Head to nerve ring	45	Head to nerve ring	46
Length vagina	90	Length spicules	33
% vulva	60	Width spicules	5
Egg diameter	33	Length tail	100
Length tail	425	Width at tail	34

posed of two anterior companion cells and four tetrad cells located adjacent to the pharyngeal tube. The intestine is modified into a trophosome or a food storage organ. In the female, the vagina is barrel-shaped, ovaries are outstretched, anus is absent. Eggs are retained in the body or deposited within the final cuticle, formed as a "capsule". The eggs are shaped like a "mushroom cap". In the male, the spicule is single, bursa and genital papillae absent, and testes outstretched. The pharynx is absent. Quantitative measurements of the adults are presented in Tables 59 and 60.

Infective stage juveniles — This stage is 188 to 250 μm long and 4 μm wide. The stylet is 9 μm long.

2. Bionomics[147-150]

Life cycle — *T. plicans* is a parasite of sciarid (mycetophilid) flies and occurs in larvae, pupae, and adult stages of these hosts. From 2 to 20 parasites of both sexes could be found in a single host. The females are five or more times the length of the males and occasionally have the peculiar characteristic of depositing their eggs beneath a layer of cuticle in the midregion (probably retaining the last juvenile cuticle) and forming what have been termed "capsules" (Figure 26). This is an unfortunate term for this condition since "capsule" has another definition related to insect immunity. However, this condition does not always exist, and many females retain the eggs in their body.

Mating and oviposition occur inside the living host, although the eggs do not hatch in that location. Development from entry to the adult stage took about 8 days, although the nematodes remained in the host for about 4 weeks and the number of eggs produced by a single female ranged up to 12,000. The egg-laden females escape from their hosts through a hole in the body cavity and deposit their eggs in the environment. When heavily infected hosts die, the eggs may be released in the body cavity of the host. The eggs embryonate while still in the female and may hatch within 24 hr after being deposited.

The actual process of penetration apparently has never been observed. Eggs deposited in the environment hatch without being ingested by potential hosts; therefore, the newly hatched juvenile is probably the infective stage. The consensus of opinion is that the infection occurs through the gut wall after the juveniles (or possibly eggs) are ingested by a host, although penetration through the cuticle is also a possibility. Nourishment uptake probably occurs through the body wall of the parasite.

In some cases, the eggs of *T. plicans* hatch while still under the cuticle of the female in the host. Under such conditions, the newly hatched juvenile could enter the host's hemocoel and initiate a second generation, although usually the host is killed by this time.

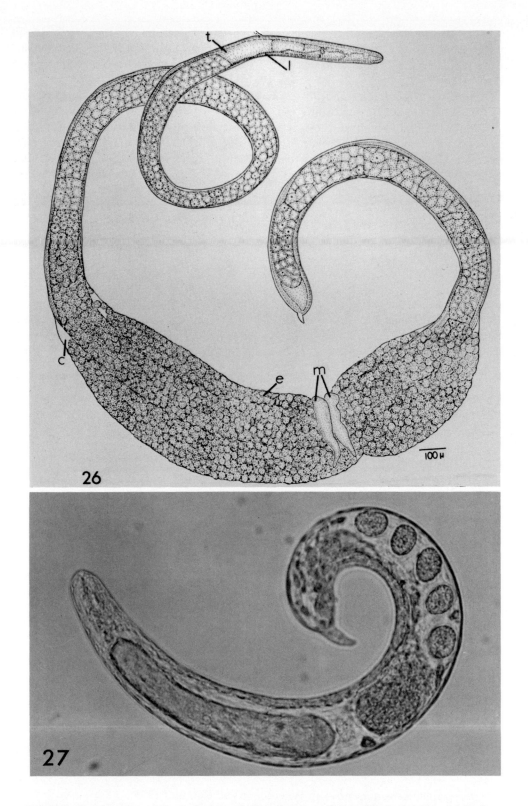

FIGURE 26 and 27. (26) A female of *Tetradonema plicans* with two males attached; c, capsule; e, eggs; l, pharyngeal canal; m, males; t, tetrad. (Courtesy of V. Ferris and J. Ferris.) (27) An ovoviviparous male of the tetradonematid *Heterogonema ovomasculis*. (Courtesy of M. Remillet.)

Eggs kept at 21°C began to hatch 15 days after oviposition. The sex ratio of the parasites was approximately three males to one female, although it may be environmentally determined like mermithids since when a single nematode was present, it was nearly always a female.

Natural occurrence — *T. plicans* probably has a world-wide distribution and has been found occurring naturally in Kansas,[148] Virginia,[149] England,[150] and California.[150a] The first three above reported infection in greenhouse pots, indicating that under greenhouse conditions the rate of parasitism increases and is more easily observed. Hungerford[148] mentioned that all attempts to locate field-infected flies were unsuccessful. Only the specimens from California were found in sciarid fly larvae removed from under the bark of a dying tree. The nematodes are distributed by infected hosts, especially the adults, and the eggs can be carried by water, or possibly wind, to new areas.

3. Host Range

Identified hosts of *T. plicans* in the U.S. and Britain belong to two species of Sciaridae, e. g., *Bradysia coprophila* (Lintner) and *Bradysia paupera* Tuom., respectively. The hosts of the California parasites were unidentified members of the same family, and it is questionable whether *T. plicans* can infect representatives of other dipterous families. Infected sciarid larvae showed a reduction of body fat and infections in 1 to 5-day-old host larvae were lethal. When the parasites were carried into the adult stage of the host, most of the latter were sterilized.

4. Culture

In vivo — It is fairly easy to maintain cultures of *T. plicans* in most soil or humans as long as noninfected hosts are periodically added to the system.

For mass production of parasites, Hudson[151] placed eggs or infective stage juveniles of *T. plicans* on the surface of pots containing 7- to 9-day-old maggots. After 12 days, infected host larvae were collected with ovipositing female parasites, and free nematode eggs were obtained by macerating whole infected maggots in a tissue homogenizer. Eggs stored in water at 10°C remained viable (less than 30%) and infective for 1 year and constituted the most suitable stage for storage. The eggs could be applied to the host substrate in water, and 100% infection was achieved against sciarids in compost by applying 1200 infective stage juveniles in 1 g of compost.[151]

In his original account of the biology of this nematode, Hungerford[148] mentioned that when potted plants containing healthy sciarids were placed in a rearing cage together with some soil containing infected maggots, the healthy maggots shortly became infected. Regarding the experimental release of *T. plicans* in greenhouses or mushroom beds, apparently no further attempts of control, following Hungerford's observations, have been conducted, although the time does seem appropriate.

B. *Heterogonema* Van Waerebeke and Remillet 1973

1. Description

The genus *Heterogonema* is characterized by the presence of two parasitic generations with the first one consisting of parthenogenetic females lacking a vulva. The subsequent generation includes females and males, but aside from producing sperm, the latter also are capable of producing eggs which hatch in their body.

C. *Heterogonema ovomasculis* van Waerebeke and Remillet 1973

This curious nematode was found parasitizing a species of nitidulid beetle in Madagascar and is capable of sterilizing the adult host. The life cycle is certainly "aberrant" since in no other known case are male nematodes capable of producing young.

TABLE 61

Measurements of the Parasitic Females of *Heterogonema ovomasculis*[152]

	Value	
Character	First generation	Second generation
Length (mm)	0.420—1.100	0.645—1.040
Greatest width	44—120	43—52
Head to nerve ring	35	25
% vulva		50
Egg diameter	20—30	23

TABLE 62

Measurements of the Male of *Heterogonema ovomasculis*[152]

Character	Value
Length (mm)	0.40—0.50
Greatest width	34—47
Head to nerve ring	26
Length spicule	33
Width spicule	2.5—5
Length tail	28—45

1. Description

First generation females — They possess little mobility and are tightly coiled into a spiral. The head is rounded and the stylet is visible. There are four large nucleated cells (tetrad) in the anterior portion of the body. Vulva and anus are absent.

Infective stage juveniles — This stage is 110 μm long and 6 μm wide. The stylet is 5 μm long. It is slender with a well-developed tetrad and an obtuse tail.

Second generation females — The body is cylindrical, with both tips rounded. There is a small stylet (3 μm) and a tetrad is present. The vulva is reduced and is used for mating, but not for oviposition. The mature eggs are discoid and surrounded by an irregular swollen membrane.

Males — With both ends rounded; the stylet is vestigal. The gonad consists of one testis and one ovary, both functional. There is a single spicule and the intestine and gubernaculum are absent.

Quantitative measurements of the adults are given in Tables 61 and 62.

2. Bionomics[152]

Life cycle — After the egg is ingested by a susceptible host, it hatches in the gut of the latter, and the emerging infective stage juvenile penetrates the gut wall and enters the host's hemocoel. Each juvenile develops into a parthenogenetic parasitic female which constitutes the first generation. Curiously, this stage does not possess a vulva, and the eggs hatch inside her body. The first generation females eventually burst open in the hemocoel but usually not until some of their juveniles have themselves developed into males and females of the second generation. The adults mate and the second generation females are oviparous and very active. After reaching maturity, they penetrate through the intestinal wall and deposit their eggs in the host's alimentary canal. The entire parasitic cycle lasts about 2 months. New infections occur when the eggs are ingested by other hosts. The males do not die after mating but are capable of producing eggs in their own body and thus establishing a third and possibly fourth generation (Figure 27).

3. Ecology

This nematode was discovered attacking an adult *Stelidota* (Nitidulidae) in Tananarive, Madagascar. The rate of parasitism varied from 30 to 50% in April and October, respectively. Up to ten first generation females could be found in a single adult beetle. Curiously, no infection was observed in host larvae or pupae. The nematode is capable of sterilizing female beetles, and the cycle can be easily maintained in the laboratory. The eggs of the second generation females, the only free-living stage of the parasite, are partially resistant and retain their infectivity for more than 2 months in the environment.

In nature, this parasite, found thus far only in Madagascar, has been recovered only from *Stelidota* sp. and a member of the Cucujdae. However, laboratory experiments showed that at least three species of the latter family and two species of Nitidulidae can serve as hosts (these hosts remain unidentified).

III. Diplogasteridae

Nematodes of the family Diplogasteridae have several types of interesting associations with insects. Since most diplogasterids occur in soil or decomposing plant or animal remains, the most common type of association with insects is phoresis. The "dauer" stage of many nematodes occurs either externally on the elytra, legs, or mouth appendages or internally in the genital chamber, trachea, or pharyngeal glands of various insects. Since members of the above-mentioned groups rarely bring about insect mortality, they will not be mentioned further here. These forms are discussed and listed elsewhere.[1]

The species of diplogasterids mentioned here are those which have been shown to cause insect disease by actively entering and reproducing within the body cavity of the insect. Since these nematodes generally kill the insect soon after they enter the body cavity, they will usually be found on dead insects. However, it should be noted that there are many microbivorous diplogasterids in the soil that can be attracted to dead insects but which never played a role in the natural mortality. For this reason, every species of Diplogasteridae found in association with insect cadavers should be cultured and then presented to healthy insects to determine if they are capable of initiating infection and disease.

The diagnostic characters of the family Diplogasteridae are listed below.

A. Diplogasteridae (Micoletzky): Rhabditida (Oerley)

A stylet is absent, and the stoma is usually broad and short with the metarhabdions bearing denticles, warts, or teeth that may be quite large. The pharynx has a median valvated bulb and a basal valveless bulb. The female has single or paired gonads and the male has paired spicules and gubernaculum. Bursa are usually small or absent. The third stage "dauer" juveniles are often covered with an oily deposit that makes them stick together or float on the surface of water.

B. *Mikoletzkya aerivora* (Cobb) 1916 (syn. *Diplogaster aerivora* Cobb syn. *Pristionchus aerivora* (Cobb))

This species has been implicated in causing mortality of several insects, and experimental evidence shows that *M. aerivora* is able to enter and destroy healthy insects.

1. Description[153]

Adults — The female has a cuticle with fine transverse and longitudinal striae. The

TABLE 63

Measurements of *Mikoletzkya aerivora* Females[153]

Character	Value
Total length (mm)	0.99—1.63
Greatest width	67—119
Length stoma	12
Length head to base of esophagus	180
Length head to nerve ring	134
% vulva	51
Length tail	195
Width at anus	39
Eggs	34 × 62

TABLE 64

Measurements of *Mikoletzkya aerivora* Males[153]

Character	Value
Total length (mm)	0.75—1.14
Greatest width	46—80
Length stoma	7.2
Length head to base of esophagus	120
Length head to nerve ring	88
Length tail	88
Width at anus	37
Length spicule	55
Width spicule	3—4
Length gubernaculum	18

head bears six fused lips, each with two papillae. The anterior papilla is extended as a seta. The base of the stoma bears a dorsal, movable, conoid tooth. A submedian projection is also present. The pharynx has a median bulb containing a valvated apparatus and a basal bulb lacking a valvated apparatus. The tail is conoid and pointed for half its length, then quickly tapers to a hairline tip. Ovaries are paired and reflexed to or beyond the vulva. Eggs develop and may hatch within the female nematode. The excretory pore is located opposite the basal bulb. In the male, the tail tapers to a fine hairline terminus. Spicules are paired, equal, slender, tapering, arcuate, and brownish, with cephalated proximal ends and acute terminals. The gubernaculum surrounds the distal portion of spicules, which are brown. There are three pairs of preanal genital papillae, two of which are ventrally submedian and one sublateral. Six pairs of postanal papillae are arranged as follows: one pair subventral immediately behind the anus, two pairs sublateral, and three closely placed pairs subventral in position. The testis is single and reflexed. Quantitative measurements are given in Tables 63 and 64. Swain[157] reported the average length of *D. aerivora* dauer stages as 350 (280 to 430) μm.

Diagnosis — The fact that the "dauer" stage juveniles (nonfeeding third stage juveniles) are associated with living insects and are capable of initiating an infection which results in host mortality biologically separates this species from other diplogasterids. Morphologically, although we know little of the variability in this species, the tapering tails of both sexes, the absence of a bursa, and the position of the genital papillae are distinguishing characters of *M. aerivora*.

2. Bionomics

Merrill and Ford[153] reported finding *M. aerivora* in the heads of the termite *Leucotermes lucifugus*. Apparently only the "dauer" or other juvenile stages were found associated with living termites since the authors stated that *M. aerivora* "never exceeded 0.50 mm in length nor completed its life cycle while within the termite." However, they go on to state that when heavy infestations occurred, the termites became sluggish and died. The authors noted that when termites were placed in moist soil with *M. aerivora,* the nematodes entered their heads (presumably in the salivary glands). After 12 days, the termites had died and their bodies were filled with nematodes.

Although Cobb described *M. aerivora* in Merrill and Ford's paper where "the nematode was associated with termites", apparently he had originally seen the same nematode previously since he remarked that *M. aerivora* was found feeding on grasshopper eggs that had been deposited in the soil at Manhattan, Kansas. Banks and Snyder[154] also reported finding juveniles of *M. aerivora* in the heads of active, normal-looking termites *Reticulitermes flavipes* and adult nematodes in sick and dead insects. A nematode that was associated with the death of *Phyllophaga* grubs was determined to be *M. aerivora,*[155] and *Heliothis obsoleta* larvae died about 8 days after they had been placed on soil containing *M. aerivora.*[156] These reports suggest the pathogenic nature of this diplogasterid nematode.

Merrill and Ford[153] stated that *M. aerivora* developed from newly hatched juvenile to adult in just 3 to 4 days in water cultures. During mating, the male wrapped itself around the female's body and the worms remained in copula for 4½ min or less. A single female that mated with seven different males laid a total of 317 fertile and 14 infertile eggs. The eggs were rapidly discharged through the vulva initially, then hatched within the mother nematode, as is typical of many rhabditids and *Neoaplectana* spp. The eggs were deposited in clusters of 6 to 30 and hatched approximately 18 hr after the first division occurred.

No specific bacteria were mentioned in association with *M. aerivora;* however, Swain[157] felt there was evidence that this and other species of *Diplogaster* are associated with some type of bacteria. Swain[157] reported *M. aerivora* as causing mortality of white-fringed beetles, *Pantomorus* sp., in the southeastern U.S. Reproducing nematodes were recovered from living beetle larvae and eventually brought about their mortality. However, Swain mentioned that it was never possible to credit *Diplogaster* with any significant control of white-fringed beetle populations.

Swain[157] cultured *D. aerivora* and other *Diplogaster* spp. on agar with the bacterium *Serratia marcescens,* as well as on beef extract agar and several other artificial media. The "dauer" juveniles could be stored in water for considerable periods and were considered the infective stages which possibly entered the mouth or anus of the host and penetrated into the body cavity.

C. *Diplogasteritus labiatus* (Cobb) 1916 syn. *Diplogaster labiata* Cobb 1916)

This species was described by Cobb in Merrill and Ford[153] from adult elm borers (*Saperda tridentata* Oliv.) in Kansas. It was considered by these authors capable of destroying healthy adult beetles and is therefore listed in this section.

1. Description[153]

Adults — The female has a cuticle with fine transverse and longitudinal striations. There are six distinct lips. The stoma is irregularly cylindroid and slightly unsymmetrical at the base, where it bears a tooth. The pharynx has a well-developed, median valvated bulb. The basal bulb has a modified valvular apparatus. The tail tapers slightly to a fine tip. Ovaries are paired, opposite, reflexed. In the males, the tail begins to taper at the anus, then narrows to a very fine terminus. Spicules are equal, with

TABLE 65

Measurements of *Diplogasteritus labiatus* Females[153]

Character	Value
Total length (mm)	0.66—0.70
Greatest width	29—30
Length stoma	8
Length head to base of esophagus	139
Length head to nerve ring	112
% vulva	59
Length tail	60
Width at anus	19
Eggs	31 × 63

TABLE 66

Measurements of *Diplogasteritus labiatus* Males[153]

Character	Value
Total length (mm)	0.60—0.72
Greatest width	20—28
Length stoma	14
Length head to base of esophagus	151
Length head to nerve ring	115
Length tail	43
Width at anus	21
Length spicule	26—32
Length gubernaculum	13—16

blunt tips, with the narrow proximal portion set off from the remainder. The gubernaculum has a rounded proximal portion. Nine pairs of papillae are present: three pairs of preanal genital papillae, one pair located adjacent to the anus, and five pairs located postanally. Quantitative characters are listed in Tables 65 and 66.

Diagnosis — Both *D. labiatus* and *M. aerivora* should be redescribed when adequate material is again collected. At this time, variation within the species should be recorded. Cobb in Merrill and Ford[153] separated the two species mainly on biological characteristics (e.g., eggs deposited singly or in groups, number of eggs hatching in body of female, manner of molting, time of mating, number of matings, ovipositional period, and location in the body of their respective insect hosts). Although these characters might well represent basic differences between the species, they can hardly be used when examining preserved specimens. Cobb did mention that the adults of *M. aerivora* were larger than those of *D. labiatus* and the spicule length of the former is about twice that of the latter. The number and arrangement of the genital papillae also differ in Cobb's description. This species was assigned to the genus *Diplogasteritus* Paramonov, whose members possess a rudimentary bursa. Although Cobb did not describe such, it is highly likely that one existed.

2. Bionomics[153]

Life cycle — The "dauer" stage juveniles of *D. labiatus* were capable of entering the intestine of adult elm beetles (*S. tridentata*) and establishing reproducing colonies. Their activities eventually ruptured the walls of the alimentary tract, and the nematodes invaded the insect's hemocoel and caused its death.

Adult females deposited their eggs singly, and hatching time in water cultures was

TABLE 67

Measurements of Female *Mesodiplogaster lheritieri*

| | Value | |
Character	After Maupas[159]	After Hirschmann[160]
Total length (mm)	1.14—1.70	1.04—1.85
Greatest width	71—128	59—121
Length stoma	10—14	
Length head to base of esophagus	185—200	143—211
Length head to excretory pore	163	
Length head to nerve ring	121	
% vulva	44—51	44—50
Length tail	230—357	189—420
Width at anus	48	

TABLE 68

Measurements of Male *Mesodiplogaster lheritieri*

| | Value | |
Character	After Maupas[159]	After Hirschmann[160]
Total length (mm)	0.87—1.22	0.87—1.26
Greatest width	50—89	49—78
Length stoma	11	
Length head to base of esophagus	182—185	167—203
Length tail	143—188	127—162
Width at anus		38
Length spicule	37—43	32—53
Length gubernaculum		16—22

from 30 to 32 hr after oviposition. The nematodes developed to the adult stage in 7 to 10 days. During the molting process, the nematode fastened its posterior end to the substrate and emerged from a break in the cuticle at the anterior end. During mating, the males wrapped themselves around the middle of the female's body. Mating lasted from 2 to 30 min. The nematodes could be grown on artificial media and therefore would be available for biological control tests.

D. *Mesodiplogaster lheritieri* (Maupas) 1919 (syn. *Diplogaster longicauda* Claus, syn. *Diplogaster lheritieri* Maupas)

Experimental studies demonstrated the ability of *M. lheritieri* to bring about insect mortality,[158] and it has been found in several habitats.

1. Description

Adults — Maupas[159] found the cuticle to be adorned with heavy longitudinal and fine transverse striae and the head surrounded by three inconspicuous lips, each of which bears two fine papillae. Males had nine pairs of genital papillae. Spicules were paired and equal. Quantitative measurements are given in Tables 67 and 68.

Hirschmann[160] described two types of mouth cavities in *M. lheritieri,* a stenostoma and a eurystoma form. The other morphological characters are identical in both forms. Hirschmann[160] adds little to the original description except to describe the mouth cavities of both forms in detail and show some variation in the position of the genital openings.

Diagnosis — The females have opposed gonads and may be oviparous or viviparous. The tails of both sexes are long and fine. Goodey[161] describes a narrow bursa as being characteristic of the genus *Mesodiplogaster;* however, neither Maupas,[159] Hirschmann,[160] or Weingartner[162] describes or figures a bursa. Weiser[163] mentions a bursa in his account of this nematode. Diagnostic characters are the structure of the stoma, the presence of ventral processes on the gubernaculum, and the position of the genital papillae, namely that one pair is located posterior to the set of three.

2. Bionomics

Life cycle — Maupas[159] described this nematode as being oviparous with the eggs hatching approximately 50 hr after being laid (at 17°C). The complete life cycle was completed in 5 days.

Pathogenicity — Weiser[163] was the first to implicate this nematode as an insect pathogen. After placing "dauer" stage juveniles (length, 200 to 300 µm) of *M. lheritieri* with *Galleria mellonella* larvae, Weiser observed mortality of the insects shortly after 24 hr. The nematodes fed and developed in the insect cadaver for 4 to 6 weeks. The dimensions of the eggs were 40 × 60 µm and the "dauer" stages were covered with an oily deposit.

At 20°C, Weiser[163] found that one generation lasted 5 days, and "dauer" or infective stages were only formed when nourishment was nearly depleted. They can survive at 34°C for up to 2 hr and at 35°C for 1 hr. They can tolerate 0.1% formalin for 12 hr, but 0.2% only for 1 hr. Weiser[163] felt that these "dauer" stages could penetrate into the body cavity of healthy insects and develop on the introduced bacteria and insect remains. In controlled experiments, Poinar[158] showed that development in the host's intestine was a prerequisite for invasion of the hemocoel. In these experiments, there was no indication that the "dauer" stages could penetrate through the intestinal wall of healthy insects without first establishing a colony inside the intestine. This is similar to the pattern of infection for *D. labiatus* reported by Merrill and Ford.[153]

3. Culture

Weiser[163] successfully cultured *M. lheritieri* on larvae of *G. mellonella* and transferred this nematode into axenic culture using agar tubes containing a piece of raw, sterile liver or kidney.

4. Host Range

Weiser[163] remarked that cultures of *M. lheritieri* were initially isolated from dead larvae of *Saperda carcharias* (L.), *Cryptorrhynchoides lapathi* (L.), *Melolontha melolontha,* and *Hoplia* sp. A complete list of hosts is given in Table 69.

5. Rate of Infection

According to Weiser,[163] 20% of the sampled population of *S. carcharias* was infected with this nematode, as well as 1 to 2% of *Melolontha* grubs and 6% of *Hoplia* grubs examined. Since this nematode grows well on a variety of artificial media, there would be no problem in obtaining adequate material for control experiments. It would probably be most successful against soil insects with biting mouthparts.

E. *Pristionchus uniformis* Fedorko and Stanuszek 1971

This nematode was found infecting different stages of *Leptinotarsa decemlineata* (Colorado potato beetle) and *Melolontha melolontha* (May bettle) in Poland.

TABLE 69

Hosts of *Mesiodiplogaster lheritieri*[163]

Order	Family	Host	Stage	Infections[a]
Coleoptera	Cerambycidae	*Saperda carcharias* (L.)	L	N
	Chrysomelidae	*Leptinotarsa decemlineata* Say	L	E
	Curculionidae	*Cryptorrhynchoides lapathi* (L.)	L	N
	Scarabaeidae	*Hoplia* sp.	L	N
		Melolontha melolontha L.	L	N
Hymenoptera	Tenthredinidae	*Neodiprion sertifer* L.	L	E
Lepidoptera	Galleriidae	*Galleria mellonella* L.	L	E
	Pieridae	*Pieris brassicae* L.	L	E
	Pyralidae	*Ephestia kuhniella* Zell.	L	E
		Pyrausta nubilalis L.	L	E
	Tortricidae	*Laspeyresia pomonella* (L.)	L	E

[a]   N, natural infection; E, experimental infection.

1. Description[164]

Adults — The head bears six partially fused lips, each bearing a terminal papilla. The stoma is shallow. A long mobile tooth (8 μm long) on the dorsal wall occupies the lumen of the metastoma. The pharynx is expanded beneath the stoma; the procorpus is narrow and expands into an enlarged valvated metacorpus. The isthmus is short and leads to the expanded nonvalvated basal bulb. The female has a cuticle with faint longitudinal and transverse striae. The tail ends in a long, thin, stiff spine. The vulva protrudes slightly, gonads are opposed, and ovaries are reflexed. In the male, the tail is bent, composed of a wide conic portion and a long, thin, stiff spine. The testis is reflexed. Spicules are paired and bent. The capitulum is separated from the body by a constriction, the distal end is pointed. The gubernaculum is short, with the tip bent backwards like a hook. The bursa is narrow and is supported by nine pairs of papillae. Quantitative measurements of the adults are given in Tables 70 and 71.

Infective stage juveniles — This stage is 242 (215 to 297) μm long and 14 (12 to 17) μm wide. The length from head to excretory pore is 15 (13 to 17) μm; the tail is 32 (27 to 38) μm long.

Diagnosis — The long tail in both sexes, the presence of a long narrow bursa, and the arrangement of the genital papillae separate *P. uniformis* from other diplogasterids associated with insects.

2. Bionomics

Life cycle — Fedorko and Stanuszek[164] found that *P. uniformis* completed one generation in 4 days at 25°C and in 9 days at 5°C (Figure 28). Four juvenile stages were present. Third stage "dauer" juveniles initiate infection and are covered with an oily deposit. They can stand on their tail and swing the body from side to side in search of an insect host. The above authors stated that the infective stage juveniles entered the intestine of *M. melolontha* grubs where they initiated development. Eventually, the developing nematodes destroyed and broke through the intestinal wall into the hemocoel. With *L. decemlineata,* the infective stages of *P. uniformis* evidently enter directly into the host's hemocoel. Using *G. mellonella* as host, Poinar[158] showed that when insect mortality occurred, it resulted from the developing nematodes breaking through the intestinal wall and entering the hemocoel. It was also stressed that certain insects may be more susceptible to diplogasterid nematodes as a result of their feeding habits, physiological condition, and the physical and chemical properties within their intestine.

TABLE 70

Measurements of Female *Pristionchus uniformis* [164]

	Value			
	From insects (N = 15)		From cultures (N = 50)	
Character	$\overline{X}$	Range	$\overline{X}$	Range
Total length (mm)	2.13	1.80—2.60	1.46	1.10—1.76
Greatest width	186	143—245	125	87—147
Length stoma	13	11—15	10	9—12
Width stoma	10	8—11	9	7—10
Length head to base of esophagus	273	207—393	244	193—322
Length head to excretory pore	104	82—156	97	73—180
Length head to nerve ring	190	157—207	173	129—270
% vulva	50	48—52	49	48—51
Length tail	312	200—357	251	229—346
Width at anus	63	53—79	42	35—64
Eggs	45 × 70		43 × 75	34-55 × 69-109

TABLE 71

Measurements of Male *Pristionchus uniformis* [164]

	Value			
	From insects (N = 10)		From cultures (N = 50)	
Character	$\overline{X}$	Range	$\overline{X}$	Range
Total length (mm)	0.69	0.58—1.13	0.76	0.63—1.35
Greatest width	52	41—57	64	48—75
Length stoma	8	6—8	8	6—10
Width stoma	5	5—7	6	4—8
Length head to base of esophagus	118	105—133	139	113—189
Length head to excretory pore	46	37—51	56	36—72
Length head to nerve ring	79	67—90	99	70—127
Length reflexion of testis	148	133—192	189	170—235
Length tail	128	107—293	131	111—305
Width at anus	38	31—86	39	33—87
Length spicule	44	37—103	46	31—124
Width spicule	5	4—17	5	4—12
Length gubernaculum	15	12—36	17	13—37
Width gubernaculum	4	3—10	4	34—11
Length bursa	130	105—299	136	115—310

3. Bacterial Associates

Sandner et al.[165] claimed to have isolated two bacterial species from the infective stage juveniles of *P. uniformis*. One was a Gram-positive coccus very close to *Streptococcus durans*. The other was a Gram-positive rod similar to *Bacillus subtilis*. The authors felt that these bacteria were carried into the host by the infective stages. It is quite probable that this is the case; however, the bacteria are probably found on the cuticle of the infective stages. There is no evidence that the bacteria are carried inside the alimentary tract of the nematode. Sandner et al.[165] surface-sterilized their infective stage nematodes with 0.1% solutions of streptomycin and penicillin. Antibiotics, however, are not very effective against bacterial spores and, therefore, might not destroy a *Bacillus* or *Streptococcus* on the cuticle of the nematode.

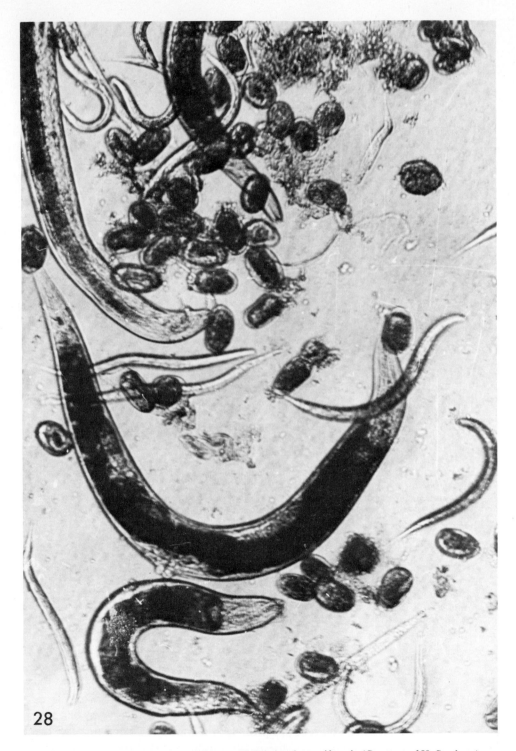

FIGURE 28.   A population of the diplogasterid *Pristionchus uniformis.* (Courtesy of H. Sandner.)

Using Hyamine 10-X solution for surface sterilizing the nematodes, the present author was unable to detect any bacteria inside the alimentary tract of the infective stage juveniles of *P. uniformis.* Poinar[158] pointed out that pathogenic bacteria adhering to the nematode's cuticle could mask any influence of the nematodes in relation to insect

TABLE 72

Infection of *Leptinotarsa decemlineata* by *Pristionchus uni-formis* at Different Seasons in Poland[166]

Season	No. of individuals infected	No. of individuals examined	% infection
Winter	1092	234	21.4
Spring	1233	72	5.8
Summer	4455	16	0.4

TABLE 73

Seasonal Variation of Infection of *Lep-tinotarsa decemlineata* with *Pristionchus uniformis*[171]

Month	% infection
January	0
February	0
March	83
April	83
May	20
June	5
July	10
August	3
September	10
October	0
November	0
December	0

diseases. Thus, at the present time, there is no conclusive evidence that *P. uniformis* or any other diplogasterid nematode possesses an infective stage juvenile that carries symbiotic bacteria in its intestine into the body cavity of an insect.

4. Culture

P. uniformis has been maintained on cultures of *G. mellonella* larvae[158] and also on solid, semiliquid, and liquid media under xenic and axenic conditions.[164] The solid medium consisted of raw liver pieces on an agar base, the semiliquid medium consisted of egg yolk poured over an agar base, and the liquid medium consisted of raw liver extract and the white of an egg in a NaCl solution. The most successful method was on solid media under xenic conditions. The nematodes only developed for a few generations on liquid media and for two generations on semiliquid media. Axenic cultures developed very slowly and never reproduced.

5. Ecology and Experimental Infections

Most infections of the Colorado potato beetle by *P. uniformis* in Poland occurred in winter and early spring just before the beetles left the soil[166] (Table 72). The nematode seemed to be fairly evenly distributed over the whole of Poland. Later results reported by Fedorko[167] showed the highest infections occurring in March and April (Table 73). Under laboratory conditions, the infection was greatest at temperatures from 0 to 13°C and from 17 to 25°C. Nematode invasion occurred at both temperatures, but development only occurred at the lower temperatures.

TABLE 74

Results of Greenhouse Experiments Using *Pristionchus uniformis* Against *Leptinotarsa decemlineata*[167]

| Experiment | Nematodes applied to | | % mortality |
	Soil	Plant	
1	+	−	38.6
2	+	+	90.5
3	−	+	86.4
Control	−	−	6.6

TABLE 75

Mortality of *Leptinotarsa decemlineata* Adults due to *Pristionchus uniformis* Applied to Soil in the Field[167]

	No. of insects insects introduced	No. of dead insects in autumn	No. of living beetles in spring
Nematode treated	2800	980 (35%)	4 (0.14%)
Control	2800	196 (7%)	353 (12.6%)

Greenhouse experiments were conducted by placing *P. uniformis* in soil with potatoes or by spraying the nematodes on the potato plants. The results are recorded in Table 74. It is interesting that although the nematodes entered and killed the insects, they failed to develop.

In contrast, Fedorko[167] reported that in another experiment, 80% of diapausing Colorado beetles were killed when their cages were sprayed with a nematode suspension (500,000 nematodes per square meter). In this case, 93% of the infected insects showed nematode development. Results of a field experiment were also reported by Fedorko[167] and are shown in Table 75. Adult beetles were allowed to enter treated soil for their winter hibernation and encountered the nematodes as they burrowed into the soil. From these experiments Fedorko[167] concluded that *P. uniformis* was adapted to low temperatures and could be used for winter control of the Colorado potato beetle.

F. Other Diplogasteridae Associated with Insect Diseases

Niklas[168] cited the nematode *Diplogasteroides berwigi* Rühm as the only parasite that played an important role in controlling *Melolontha* larvae in Germany. The nematodes were present during the entire developmental period of the insects, but high mortality levels occurred in May and November. The possibility of this nematode serving as a vector of infectious microorganisms was also mentioned.

Aside from *M. aerivora,* Swain[169] also briefly discussed other unidentified species of Diplogasteridae that could cause mortality in white-fringed beetle larvae (*Pantomorus* sp.). Clearly, more research with this group of nematodes is necessary to evaluate their effectiveness as insect pathogens.

IV. Rhabditidae

Most representatives of this family are free-living; however, many of these have

TABLE 76

Measurements of *Rhabditis insectivora* Females[170]

	Value	
Character	Free-living (N = 5)	From insect (N = 5)
Total length (mm)	2.20 — 2.91	1.98—3.21
Greatest width	125 — 169	134—189
Length head to base of esophagus	268	—
Length head to excretory pore	225	—
Length head to nerve ring	175	—
% vulva	47 — 51	49—56
Length tail	150	—
Width at anus	30	—

various types of phoretic relationships with insects. Since such relationships are generally of no serious consequence to the invertebrate, they will not be discussed here.

Some rhabditids are able to live inside the intestine or reproductive system of insects. Thus *Oryctonema genitalis* and *Eudronema intestinalis* develop in the bursa copulatrix and intestine, respectively, of certain beetles.

Although there are several accounts of rhabditid nematodes acting as facultative parasites or pathogens, most reports are vague and only one species is described in detail here. Other rhabditids known to bring about insect mortality are also discussed since some of these species could be cultured on artificial media and possibly used in a biological control program.

A. *Rhabditis insectivora* Körner 1954

This species was described as a facultative parasite of the cerambycid beetle *Dorcus parallelopipedus* in Germany. It is essentially a free-living rhabditid whose "dauer" stages have the ability to enter healthy insects and bring about their mortality. Körner[170] has been the only one to report on the presence of this nematode.

1. Description[170]

Adults — The six lips are slightly offset, each bearing two short papillae. The stoma is short, with glottoid apparatus bearing three small teeth. The pharynx is cylindrical, with a pronounced isthmus and an enlarged basal bulb with valves. The excretory pore is located opposite the basal bulb. The tail is pointed. In the female, the vulva is median, gonads paired and reflexed, and the tail conical and pointed. The male has a leptoderan bursa, with nine pairs of bursal papillae and a small pair of phasmids shaped like elongated papillae. Spicules are paired and separate. A gubernaculum is present. Quantitative measurements are given in Tables 76 and 77.

"**Dauer**" **juveniles** — The mouth is closed. This stage is 666 to 725 μm long and 30 to 35 μm wide.

2. Bionomics[170]

R. insectivora is capable of continuous multiplication in the host's environment; thus, an insect is not essential to the development of this nematode. However, "dauer" juveniles are formed which are capable of entering the intestine and reaching the hemocoel of *D. parallelopipedus* larvae. In this location, the nematodes develop and mate, but most females are apparently incapable of depositing their eggs inside the beetle. When the infection rates are high, the beetles can be killed. About 60% of the beetle larvae were infected in one locality.

TABLE 77

Measurements of *Rhabditis insectivora* Males[170]

| Character | Value | |
	Free-living (N = 5)	From insect (N = 5)
Total length (mm)	25	—
Greatest width	78—91	76—159
Length tail	25	—
Length spicule	88—95	85—118
Length gubernaculum	37—51	38—54

The "dauer" juveniles would develop on beetle feces or dead beetle larvae as well as enter healthy beetle larvae in laboratory experiments. If the infected beetle larvae died, the nematodes could then continue their development in the cadaver. This suggests that the inhibition of nematode development inside the host was due to a nutritional inadequacy which was removed when microorganisms invaded the dead insects.

R. insectivora was recovered from only the above-mentioned cerambycid beetle and has not been mentioned since Körner's report as causing insect mortality. Apparently, it has never been used as a biological control agent.

B. Other *Rhabditis* sp. Causing Insect Disease

Griffith[171] reported on a species of *Rhabditis* that occurred in the decomposing tissue of coconut palms in Trinidad. Unfortunately, a description of the nematode was not presented so its true identity is not known. However, Griffith claimed that the nematodes were capable of entering the body cavity of larvae of the palm weevil (*Rhynchophorus palmarum*) and killing some of the insects. This relationship should be investigated further and controlled experiments conducted to elucidate the life cycle.

Mizuta and Sato[172] recently discovered a *Rhabditis* sp. causing mortality of silkworms (*Bombyx mori*) in Japan. The nematodes were thought to be ingested after they were splashed onto mulberry leaves during the rainy season. The nematodes then penetrated the intestinal wall and entered the hemocoel. Infected silkworm larvae lost their appetite, became weak, and finally died.

Another species of *Rhabditis* was recovered from the larvae of rhinoceros beetles on the west coast of Ceylon.[173] Laboratory experiments showed that this nematode could actively enter the hemocoel of *Oryctes* larvae, pupae, or adults and cause their death in 7 to 21 days. The "dauer" stages of this nematode were clustered around the spiracles and within the trachea of living beetles and could have entered the body cavity by their openings. This rhabditid was sent to Fiji where it was reared on artificial media for several years, then released in the field against *O. rhinoceros* in Western Samoa, American Samoa, Wallis Island, and Fiji. It was recovered from the field at least once in Fiji, but further information concerning these releases is unavailable.

Kurian[174] reared another rhabditid on a potato agar bacterial medium to obtain material for infections against *O. rhinoceros* larvae. Introduction of various stages of this nematode to third instar beetle larvae produced mortality within 14 days, at which time nematodes were recovered from the body cavity of the grubs.

Other reports (cited by Poinar)[175] concerning the presence of *Rhabditis* spp. in field-collected insects also exist and suggest that certain species of free-living rhabditids have the ability to enter and destroy healthy insects. Although the host range of these nematodes may be rather broad, it would be no difficult problem to culture them on artificial media, harvest the "dauer" stages, and release them for control experiments.

Care should be taken to insure that the habitat at the release site is not one which stimulates the "dauer" to initiate development in the absence of an insect host.

Nematodes of the genus *Parasitorhabditis* commonly inhabit the frass of bark- or wood-eating beetles and have one or more stages that occur in the intestine of the beetles. Some damage has been reported to the intestinal cells when nematode populations build up and some species of *Parasitorhabditis* actively enter the hemocoel of their hosts. In the latter case, this action may bring about insect mortality. These cases are summarized by Poinar.[175] Since then, Blinova and Gurando[176] described *P. fuchsi* which develops to the adult stages in the hemocoel of the scolytid beetle *Blastophagus minor.*

V. Steinernematidae

The family Steinernematidae, with *Steinernema Travassos* as the type genus, was defined by Chitwood and Chitwood[177,178] as rhabditoid forms with indistinct lips, reduced stoma, reduced pharyngeal bulb, and males without a bursa. Further characterization was presented by Turco et al.[179] and Poinar.[1] The genus *Neoaplectana* was established by Steiner,[180] whereas Filipjev,[181] noting the resemblance between *Neoaplectana* and *Steinernema,* first placed both genera in the subfamily Steinernematinae. This subfamily was then erected to the family level by Chitwood and Chitwood.[177,178] Skrjabin et al.[182] considered *Steinernema* as a junior synonym of *Oxysomatium,* thus resulting in the family name Neoaplectanidae. However, this synonymy has been rejected by most workers and is not accepted here.

Members of the genus *Neoaplectana* have been extensively studied and used as biological control agents on numerous occasions. However, literature on the systematics and biology of this genus has never been collected together in summary form, and as a result, it has been difficult to get an overview of the entire group. For that reason, a review of the entire genus is presented in this work, including information on systematics, biology, host-parasite relationship, and the associated bacteria. Aside from members of the genus *Heterorhabditis,* no other known group of nematodes has a symbiotic relationship with bacteria.

The neoaplectanid nematodes undoubtedly arose from free-living microbivorous soil nematodes (Rhabditoidea) and are not really that much removed from these less specialized groups today. The main differences that arose in these insect parasites were (1) the development of a "dauer" stage that became infective for insects and could actively enter natural openings of the latter; (2) the acquisition of a specific bacterium that had the "right" characteristics, so could develop along with the nematode and not overrun or inhibit its growth, and which "possessed" the ability to remain in the intestine of the nematode during the formation of the infective stage; and (3) the ability of the females (especially of the first generation) to produce large numbers of eggs for quick colonization of the insect cadaver before other soil agents could enter.

Morphological changes resulting from insect parasitism are also apparent. Living in a semiliquid environment resulted in the reduction of the valvular apparatus of the basal bulb. Another change is the extreme length (relatively speaking) of the first generation adults (especially females) resulting from an ample supply of nutrients inside the insect host.

A. Systematics: Steinernematidae Chitwood and Chitwood, Rhabditoidea (Oerley) Rhabditida (Oerley) (syn. Neoaplectanidae Sobolev)

A stylet is absent. Six lips are partially or completely fused and amphids are present on the lateral lips. The pharynx is composed of a cylindrical procorpus, a slightly swollen metacorpus, and a basal bulb. The bulb contains a modified valve with only refractile ridges lining the walls of the lumen. The ovaries are paired and opposite and the testis is single. The spicules are paired and separate. A bursa is absent, but the gubernaculum and genital papillae are present. All are obligate parasites of insects and are associated with specific symbiotic bacteria which provide nutrients to the developing forms.

Several factors make it difficult to diagnose species in this family. The first is that many of the early descriptions failed to discuss certain characters which are now considered important in characterizing these forms. The second is the extreme variability within any one population of these nematodes. Consider the variability, both quantitative and qualitative, found within members of the species *Neoplectana glaseri*. Not only do the lengths and other measurements of the first and second generation adults differ significantly, but the female tail shape is strikingly different. Such variability is not found only in *N. glaseri* or *N. carpocapsae*, as pointed out by Stanuszek,[183,184] but also occurs within the other described species. Unfortunately, most of the descriptions of neoaplectanid nematodes do not give a good indication of the variability of characters.

Shape of the spicules and gubernaculum, body length, and body width all vary greatly within populations and even overlap between species. Thus, the characters that Turco et al.[179] used in their key to differentiate between the species of *Neoaplectana* are considered by the present author too variable to be of any value.

What characters, then, can be used to separate species in this family? One of the least variable, and perhaps the most reliable, is the length of the infective stage juveniles. Their formation occurs at a certain time in the development of the nematode, a time when nourishment is fairly depleted and therefore does not have much of an effect on size. Also, when ample nourishment is available, it is the adults which demonstrate its effect by continuing to grow after maturity. This is especially evident in female neoaplectanids that enter a healthy insect.

A second character that seems to have some reliability is the spine at the tip of the male tail. This spine is always absent in *N. glaseri* and absent in most males of *N. bibionis*. However, it does appear on most males of *N. carpocapsae* and *N. feltiae*. The size of the spine is more difficult to judge, since few studies have been conducted on the variability of that character.

The third character that appears to be fairly stable is the distance of the excretory pore from the head end of the adult nematode. It is obvious that the excretory pore is posteriorly placed in *N. glaseri* in comparison with *N. carpocapsae*, and this character is represented here by the ratio formed by dividing the distance of the excretory pore from the head by the length of the pharynx. This ratio does vary somewhat between male and female and first and second generation but is still consistent enough to show differences between some of the species.

While in the Soviet Union, the present author examined preserved specimens of *N. arenaria* Artyukhovsky 1967, *N. kirjanovae* Veremchuk 1966, *N. belorussica* Veremchuk 1966, *N. bothynoderi* Kirjanova and Puchkova 1955, *N. georgica* Kakulia and Veremchuk, 1965, and *N. elateridicola* Veremchuk 1970. A redescription of *N. bibionis* is provided in the present work from specimens the author collected earlier.[185] Natural populations of *N. glaseri* were obtained from North Carolina[186] and were redescribed.[187] The original type material of *N. hoptha* Turco and *Steinernema kraussei* Steiner was obtained from Washington, D. C. through the courtesy of Dr. Morgan Golden. The author was unable to obtain specimens of *N. menozzii* Travassos, *N. semiothisae* Veremchuk, *N. leucaniae* Hoy, or *N. melolonthae* Weiser.

1. Hybridization Studies

The biological concept of species based on genetic compatibility as expressed by

TABLE 78

Hybridization Studies with Strains and Species of *Neoaplectana*

| | *N. carpocapsae* | | | | | |
	DD-136 strain	Czechoslovakian strain	Mexican strain	Agriotos strain	*N. bibionis*	*N. glaseri*
N. carpocapsae						
DD-136 strain	+	+	+	+	-	-
Czechoslovakian strain	+	+	+	+	-	-
Mexican strain	+	+	+	+	-	-
Agriotos strain	+	+	+	+	-	-
N. bibionis	-	-	-	-	+	-
N. glaseri	-	-	-	-	-	+

Note: +, juveniles produced which in turn produced fertile adults; -, juveniles not produced.

hybridization under laboratory conditions can be applied to the neoaplectanid nematodes. Indeed, with nematodes that show so much variation within populations and have relatively few easily recognizable differences, such a species criterion can be very useful.

This criterion has already been used in defining the relationship between two populations of neoaplectanids recovered from codling moth larvae in widely separated geographical areas.[188] It was also used in determining the relationship between a neoaplectanid collected in the Soviet Union and the DD-136 strain of *N. carpocapsae*[189] and nematodes collected in Poland.[183]

In conducting hybridization studies with strains and species of neoaplectanids (Table 78), the present author individually reared the nematodes from infective stages to adults in hanging drops of *Galleria mellonella* blood. Pairs were then transferred to a new drop of blood and mating, if it occurred, usually took place within 24 hr.

Using the characters mentioned above and the results of the hybridization experiments (which unfortunately could not be used with all species because of the lack of viable material), the following key to the species of *Neoaplectana,* as recognized by the present author, was constructed.

Key to the Species of *Neoaplectana*

1. Tail tip of male lacking a spine (some exceptions may occur, but never in over 20% of the total population) — 2
1. Tail tip of male with spine (spike, projection, mucron) — 5
2. Infective stages range from 864 to 1448 μm in length; ratio of EP/pharynx ranges from 0.60 to 0.90 — *N. glaseri* Steiner 1929 (syn. *N. arenaria* Artyukkovsky 1967)
2. Infective stages less than 850 μm in length; ratio of EP/pharynx ranges from 0.20 to 0.70 — 3
3. Infective stages range between 600 and 850 μm in length — 4
3. Infective stages range between 400 and 600 μm in length — *N. menozzii* Travassos 1932.
4. Ratio of EP/pharynx ranges from 0.50 to 0.70 — *N. bibionis* Bovien 1937 (syn. *N. affinis* Bovien 1937).
4. Ratio of EP/pharynx ranges from 0.20 to 0.50 — *N. kirjanovae* Veremchuk 1966.
5. Infective stages range from 725 to 900 μm in length; spine on male tail ranges from 5 to 13 μm in length — 6

5. Infective stages range from 425 to 700 μm in length; spine on male tail ranges from 1 to 5 μm in length — *N. carpocapsae* Weiser 1955 (syn. *N. belorussica* Veremchuk 1966; *N. semiothisae* Veremchuk and Litvinchuk 1971; *N. dutkyi* Turco et al. 1971; *N. dutkii* Jackson 1965; *N. dutkii* Welch 1963)
6. Ratio of EP/pharynx over 0.75 — *N. feltiae* Filepjev 1934 (syn. *N. bothynoderi* Kirjanova and Puchkova 1955)
6. Ratio of EP/pharynx ranges from 0.60 to 0.65 — *N. georgica* Kakulia and Veremchuk 1965.

Omitted Species

1. *N. leucaniae* Hoy 1954; *N. chreisma* Steiner in Glaser, McCoy, and Girth, 1942 (also in Turco et al.[179]); *N. janckii* Weiser and Koehler 1955 — Description is insufficient for any type of diagnosis
2. *N. elateridicola* Veremchuk 1970 (syn. *N. titovi* Veremchuk 1966) — No description
4. *N. melolonthae* Weiser 1959 — Member of the Diplogasteridae and does not belong in the genus *Neoaplectana*
5. *N. hoptha* Turco 1970 — Belongs in the genus *Heterorhabditis*

2. Bionomics

In the neoaplectanids, infection of an insect host can only be initiated by the infective stage (in nature). Only these stages, which correspond to the "dauer" of the free-living nematodes, have the ability to seek out openings in an insect and actively enter the body cavity. The fact that entry can be "active" has been demonstrated several times by showing that immobile or nonfeeding stages of insects (pupae-adult Lepidoptera) can be infected by *Neoaplectana* spp. The nematodes can also be passively ingested with food, so that just about any action that brings the nematodes in close contact with the insect can be successful for infection.

The primary objective of the infective stage juvenile is to reach the body cavity of the potential host. This stage lacks a stylet, hook, or any other object that would be necessary for penetration through the insect's cuticle, and there is no evidence that the infective stages can gain entrance to the body cavity by boring through the host's body wall. Earlier assumptions of this occurring did not consider the ability of the "dauer" to enter the spiracles of the host. Thus, only natural openings can be used by the nematodes to enter the insect. These openings can be the mouth, anus, spiracles, and, of course, wounds.

It is more difficult to determine just how the nematode burrows through the gut wall in the case of entry through the intestine. Since the infective stage is a nonfeeding form, there is little way salivary or other secretions could pass out of its mouth to aid penetration, and it appears that simple, mechanical pressure is the only way these nematodes can enter their hosts. This may be why gut penetration only occurs through the midgut wall.

Even after having lost their ensheathing cuticle, the infective stages do not initiate development until they are inside the insect's body cavity. Indeed, there have been many misconceptions concerning the enclosing second stage cuticle that surrounds the third stage infective neoaplectanids. It was once thought that when the surrounding cuticle was lost, the nematodes were no longer infective. This is not the case. Many infective stages lose their cuticle in storage, or when handled roughly, and not in the crop or midgut of the host. Yet they are still infective and can continue normal development. Another misconception is that the ensheathing cuticle protects the nematode from desiccation or chemicals. Such may be the case for free-living rhabditid nematodes that "encyst" in soil or on the external surface of insects, but the outer cuticle in *Neoaplectana* spp. is relatively thin and does not really serve to protect the nematode. The infective stage is resistant to environmental conditions and chemicals because of its morphological and physiological state and not as a result of the ensheathing cuticle. Those nematodes which still retain their cuticle after entering the insect lose it in the intestinal tract before penetration. Newly penetrated infective stages removed from the host's hemocoel never contain their ensheathing cuticle.

The nematodes initiate development as soon as they enter the host's hemocoel. The alimentary tract opens and hemolymph is ingested. The nematodes develop into parasitic third stage juveniles and, within a few hours, molt to the fourth stage. As soon as the nematodes initiate development, contents of the infective stages' alimentary tract are passed into the environment — in this case, the insect hemolymph. This is how cells of the symbiotic bacteria, which are lodged in the intestinal lumen, reach the hemocoel of the insect. These cells are nonmotile and in a compact pellet inside the nematodes, but become motile soon after reaching the insects' blood. They multiply and spread through the entire body cavity of the host and establish favorable conditions for nematode development.

Sexes are separate in neoaplectanid nematodes, and each infective stage develops into a male or female. These adults which arise from the infective stages are called first generation adults and, depending on the size of the host, are usually much larger than the adults of subsequent generations. By the time the adult nematodes are formed (3 to 5 days after entry of the infective stages), the host is dead and the nematodes continue feeding on the cadaver. The males may develop a little faster than the females, and mating occurs immediately after the adults are formed. In mating of *N. carpocapsae,* the males, which are about one third to one eighth the size of the females, crawl over and around the female, which usually continues feeding during the whole process. The male continues to curl his tail over or next to the female as if he were investigating her surface. During mating, the male wraps his body perpendicularly around the female, and then after locating the vulva, inserts his spicules into the short vagina. Copulation may continue up to 1 hr, during which time sperm and spermatocytes are transferred into the uterus of the female.

Egg development is very rapid, and fertile eggs are deposited by females in 24 to 48 hr after mating. The females first deposit eggs in the host's hemocoel; later, egg development appears to overwhelm the ovipositional process or else the vaginal muscles cannot stand up under the large number of eggs, and hatching occurs inside the mother's uterus.

Just before hatching, the young first stage juvenile begins turning continuously around within the egg and forces its head against the transparent, elastic, smooth egg shell. The pharyngeal bulb can be seen pulsating, and the intestinal cells contain birefrigent bodies (urate crystals?). The nematode keeps forcing its head against the shell membrane until it breaks, and then the worm quickly exits. There is no cast cuticle left in the egg shell.

There are two alternative developmental patterns for each newly hatched juvenile. It can continue its parasitic development with four molts to the adult stage or it can develop into an infective stage juvenile. The course of development seems to depend directly on the amount of nourishment available, which in turn depends upon environmental conditions, crowding, associated microflora, etc. Those juveniles that complete their development form second generation adults which are invariably smaller than the first generation adults. Aside from the obvious quantitative differences, there may be distinct qualitative differences which separate adults of the two generations.

These second generation adults mate and produce a smaller number of eggs. Because of the crowded conditions now present in the host cadaver, most or all of the juveniles hatching from eggs deposited by second generation females will develop into infective stage juveniles. Depending on conditions or the size of the host, several parasitic generations could occur in the cadaver, each one with succeedingly smaller adults. The large sized adults so characteristic of the first generation in insects are usually not seen when the nematodes are artificially cultured.

Those first and second generation juveniles that develop into infective stages receive their stimuli to do so in the early portion of the second stage. There is a point reached when a second stage juvenile cannot form an infective third stage, even if it is removed from the insect and held in tap water.

The unique aspect of infective stage formation in neoaplectanid nematodes is the ability of the juveniles to retain the associated bacteria in their intestine. At this time in their development, the second stage juveniles (along with other stages) are ingesting not only the symbiotic bacterium, but other bacteria that have now become associated with breakdown products of the insect. During "dauer" formation, however, only the symbiotic bacterium is retained in the gut lumen of the infective stage juveniles. All others are either released through the anus or destroyed inside the nematode's intestine. After the infective stage juveniles are formed, they leave the insect cadaver and enter the environment.

There are some abnormal developmental patterns which appear in neoaplectanid nematodes. One of these is the formation of pigmy females. Pigmy females are short, stubby, swollen females that are commonly formed in some hosts in nature. The present author feels that under certain circumstances (suboptimum development) all species of *Neoaplectana* will form pigmy females. These females can produce a limited number of eggs although they certainly are not very prolific. Somewhat rarer is a condition which we have termed ballooning. Evidently, this occurs when the molt of a juvenile is not complete, and the cuticle of the head and pharyngeal lining swells up to form a balloon-like structure attached to the nematode's mouth. Such nematodes cannot feed and eventually starve.

Sometimes the infective stages were not able to molt normally. Usually there is a break in the second stage cuticle that occurs in the neck region of the nematode. The anterior end comes off like a cap, and the nematode crawls out of the posterior portion. Sometimes the posterior portion does not come off, and as the nematodes begin growing, the anterior ring of the retaining cuticle constricts their necks, eventually killing them.

As soon as the infective stages are formed, they enter the environment, and then our knowledge of their habits is very limited. Aside from being found in the soil, their movements have been little studied. The infective stages of some species (*N. carpocapsae,* for instance) appear to migrate to the soil surface, whereas others (*N. glaseri*) appear to remain beneath the soil surface.

Development within the host also varies among species. At 22°C, *N. glaseri* infectives can enter a host and develop to first generation adults in just 3 days, whereas *N. bibionis* will reach that stage in the same host in only 6 to 7 days. Their associated bacteria allow neoaplectanid nematodes to develop in just about all insects they can enter, providing they can reach the hemocoel. This is obvious by examining the host list of *N. carpocapsae.*

B. *Steinernema* Travassos 1927

In 1923, Steiner[190] described the entomogenous nematode *Aplectana kraussei* that was parasitizing larvae of the pamphilid sawfly *Cephaleia abietis* (L.) near Eberswalde, now in East Germany. In 1927, Travassos[191] erected a new genus, *Steinernema,* for this species, and Chitwood and Chitwood[177] erected the family Steinernematidae for *Steinernema* with *Oxysomatium* Raillet and Henry. This is why Sobolev[192] considered

TABLE 79

Measurements of Female *Steinernema kraussei* (Steiner)[190]

Character	Value
Total length (mm)	1.15
Greatest width	113
Length stoma	4
Length head to base of esophagus	150
Length head to excretory pore	49
Length head to nerve ring	84
% vulva	51.8
Length tail	31
Width at anus	42

Neoaplectanidae as the correct family name. However, the synonomy is not accepted here, and the present author considers *Steinernema* as a valid genus.

When Steiner erected the genus *Neoaplectana* in 1929, he separated the two genera by the following characters. "The general shape of the body, but especially the spicula and the gubernaculum, are almost the same, yet the number and arrangement of the head sense organs are very different, *Steinernema* having but a single circle of four submedial papillae, whereas *Neoaplectana* has two circles of six each. In addition, the number of male copulatory papillae is much larger in *Neoaplectana* and their arrangement is very different."[195]

Unfortunately, Steiner never did describe completely the tail papillae of *S. kraussei* and his drawings do not show a clear difference in the arrangement of the papillae of *N. glaseri* and *S. kraussei*. Thus, the only character that separates the two genera is the number and arrangement of head papillae. It was rather surprising that in his redescription of *S. kraussei*, Mraček[193] never mentioned this character.

Aside from the original description by Steiner,[190] Mraček[192] and Weiser[194] are the only ones who claim to have reisolated and studied *S. kraussei*, the only species in the genus.

C. *Steinernema kraussei* Steiner 1923
1. Description
Adults — According to Steiner,[190] the valve in the terminal pharyngeal bulb is absent. The cuticle is smooth and the lips indistinct, possibly three united. The adult bears four submedial papillae and amphids. The excretory pore is located anterior to the nerve ring. The females are variable in size. The tail tip is rounded in old females and conically tapered in younger forms. The ovaries are reflexed at the tip, and eggs develop inside the female uterus. In the male, spicules are variable in shape. The single testis is reflexed at tip. The tail papillae have a single preanal papilla and two paired postanal pairs. The tail tip bears a small projection or mucron. Quantitative data are presented in Tables 79 and 80.

According to Mraček,[193] based on material collected from *Cephaleia abietis* in Czechoslovakia, females are variable in size, with the first generation females two or three times larger than those of the second generation. First generation females produce approximately 500 eggs, many of which hatch in her body. The tail is variable, ranging from conical with a small spine to rounded without a spine. The oral opening is surrounded by a circle of rounded lips, each bearing a short seta. The terminal pharyngeal bulb is without a well-differentiated valve. Ovaries are paired and reflexed. In the male, the tail is blunt, ending in a short spine. Tail papillae vary in number and

TABLE 80

Measurements of Male *Steinernema kraussei* (Steiner)[190]

Character	Value		
Total length (mm)		1.01	
Greatest width		109	
Length head to base of esophagus		130	
Length head to excretory pore		63	
Length tail		30	
Width at anus		42	
Length spicule	50	—	54
Length gubernaculum	38	—	42

TABLE 81

Measurements of Female *Steinernema kraussei* (Steiner)[193]

	Value			
	First generation		Second generation	
Character (N = 10)	$\overline{X}$	Range	$\overline{X}$	Range
Total length (mm)	2.36	2.18—3.00	0.99	(0.81—1.71)
Greatest width	156		133	
Length stoma	7		5	(4—7)
Length head to base of esophagus	184		162	
Length head to nerve ring	138		105	
% vulva	53		58	
Length tail	42		28	
Width at anus	52		38	

TABLE 82

Measurements of Male *Steinernema kraussei* (Steiner)[193]

	Value			
	First generation		Second generation	
Character (N = 10)	$\overline{X}$	Range	$\overline{X}$	Range
Total length (mm)	1.31	(1.22—1.46)	0.99	(0.87—1.13)
Greatest width	82		79	
Length stoma	4	(3—4)	4	(3—4)
Length head to base of esophagus	165		157	
Length head to nerve ring	123		115	
Length tail	42		27	
Width at anus	32		29	
Length spicule				(48—55)
Length gubernaculum				(35—43)

development among individuals, with five or six preanal papillae and two or three postanal papillae. Spicules have a ventral arch. The gubernaculum is variable in shape. Quantitative measurements are presented in Tables 81 and 82.

Infective stage juveniles — This stage is 630 to 1050 μm long.

Diagnosis — In discussing differences between *Steinernema* and *Neoaplectana* Mraček[193] mentioned that the giant females and males of *Steinernema* are smaller than those of *Neoaplectana* and that adults of *Steinernema* possess three lips each equipped with a short seta, arranged in a crown around the mouth opening.

2. Bionomics

Mraček[193] pointed out that, in actuality, it was *S. kraussei* and not *N. janickii* that Weiser[194] discussed. Weiser mentioned that the nematode was present in 10 to 15% of the nymphs of *C. abietis* during most of the year and, that at times, 100% of the larvae were infected. Low temperatures and clay soil were conditions which limited infection.

3. Bacterial Associates

There is no mention made of bacterial associates in the original description of *S. kraussei,*[190] however, Weiser[194] mentioned that a single bacterium was isolated from host cadavers and grown on artificial media. Mraček[193] mentioned that the bacterium, which can be easily grown on standard media, was identified as a *Flavobacterium* sp. This would place it very close to *Achromobacter nematophilus* which is associated with neoaplectanid nematodes. Sawfly larvae infested with *S. kraussei* changed in color from green to gray and brown and were flabby to the touch. Aside from the natural sawfly host, only larvae of *Galleria mellonella* have served as an experimental host for this nematode.[194]

D. *Neoaplectana* Steiner 1929

This genus contains some of the most important nematode parasites of terrestrial insects and offers a tremendous potential for biological control which has never been realized. There are undoubtedly many more undiscovered species of *Neoaplectana* than are mentioned in this review, and they in turn may have characteristics that could be used to advantage in controlling insects in particular ecological niches. What is needed is a detailed analysis of the behavior and tolerance limits of each species or strain of *Neoaplectana* in order to wisely use them against select insect pests.

The first described species in this genus was *N. glaseri*, which was discovered parasitizing Japanese beetle grubs in the eastern U.S. After a relatively short but illustrious history, the species fell into oblivion as far as biological control was concerned. Recently, it has been rediscovered, and populations are now available for control studies. We can hope that, with what knowledge we have acquired during the past years, our efforts with this species will be more successful and permanent than before.

E. *Neoaplectana glaseri* Steiner 1929

1. Description

Adults — According to Steiner,[195] the tail of the female is short and conical with a blunt end; the tail of the male is broad and obtuse. The cuticle is thin, without annulations. The head is not offset. There are three indistinct lips, each with two papillae. Amphids are shifted dorsad to the same level as the lateral papillae. A buccal cavity is absent. The anterior portion of the pharynx is cylindroid and connected with the terminal bulb with a faint isthmus. Ribbed valvulae in the terminal bulb are indistinct. The ovaries are reflexed. Eggs and various stages of juveniles are found in the uterus. The testis is single and outstretched. The spicules are large and arcuate, with the tip slightly cephalated, forming a hook. The gubernaculum is large, with the distal part lineate and the proximal part broadly swollen. There is a single median and 11 to 13 ventrosubmedian, with lateral and dorsosubmedian papillae on each side. Quantitative values of adults are presented in Tables 83 and 84.

According to Poinar,[187] the adult cuticle is smooth. The head is truncate to slightly rounded, sometimes slightly offset from the rest of the body in males. The lips are united, containing an inner set of six labial papillae and an outer circlet of six cephalic

TABLE 83

Measurements of *Neoaplectana glaseri* Males[195]

Character	Value
Total length (mm)	1.4
Greatest width	93
Length head to base of esophagus	234
Length head to nerve ring	133
Length tail	47
Width at anus	62

TABLE 84

Measurements of *Neoaplectana glaseri* Females[195]

Character	Value
Total length (mm)	4.7
Greatest width	207
Length head to base of esophagus	183
Length head to nerve ring	108
% vulva	51.5
Length tail	56
Width at anus	66

papillae. Amphids are small and porelike, located just behind the circlet of six cephalic papillae. The stoma is partially collapsed. The pharyngeal collar is lacking, and the pharynx extends nearly to the mouth opening. Cheilorhabdions are represented by a darkly sclerotized zone lining the lip region just inside the mouth; beneath this is another sclerotized area that probably represents the modified prorhabdions: meso-, meta-, and telorhabdions are vestigial, although sometimes rarely represented as refractive edges lining the collapsed walls of the stoma. The pharynx is muscular, and the anterior portion of the procorpus may be slightly expanded just behind the vestibule, then extends into the slightly enlarged nonvalvated metacorpus, followed by an isthmus and terminates in a basal bulb containing a cavity lined with refractive ridges. The base of the pharynx may be inserted into the anterior portion of the intestine. The nerve ring surrounds the basal bulb or the isthmus just anterior to the basal bulb. The excretory pore is usually anterior to the nerve ring except in first generation males. Lateral fields and phasmids are inconspicuous. Females are amphidelphic with opposed ovaries that are usually reflexed; however, the ovaries are occasionally extended, especially in second generation females, where the tip of the ovary is often swollen and constricted off from the remainder. The vulva is a transverse slit, generally protruding from the body surface. The vagina is short with muscular walls leading into the uterus. Eggs commonly develop and hatch inside the reproductive system of the females, especially those of the first generation. The female tail often has a postanal swelling, terminating with a rounded projection in the first generation and a fine micron in the second generation. Pigmy forms or swollen miniature females were never found. Males have a single reflexed testis consisting of a germinal growth zone leading into a seminal vesicle containing spermatophores. Vas deferens is conspicuous with muscular walls. Spicules are paired, symmetrical, curved, and bear a rounded inconspicuous arch on their ventral surfaces. The shape of the capitulum is variable, from slightly pointed to round or flat. The surface and edge of the calamus and lamina bear ridges. The gubernaculum is variable in shape, ranging from completely flattened to bow-shaped in lateral view, with a small upturned proximal portion. Distal tips of the spicules have a concavity, giving the tips a "hooked" or "notched" appearance. The male tail has a complement of 23 anal papillae (11 pairs and a single ventral adanal) that are consistently present with little variation in position; comprising two rows of seven ventrolateral papillae, three paired postanal papillae, one pair of dorsal lateral papillae, and a single ventral adanal papilla. The tip of the tail is rounded without a mucron. A bursa is absent. Quantitative measurements of the first and second generation adults of *N. glaseri* are presented in Tables 85 and 86.

Infective stage juveniles — Third stage juveniles are enclosed in second stage cuticles. Infective stage juveniles are much narrower than the corresponding parasitic juveniles. The mouth and anal openings are closed, and the pharynx and intestine are collapsed. The tail is pointed and the lateral fields are composed of six incisures. The

TABLE 85

Comparative Measurements of First and Second Generation Females of *Neoaplectana glaseri*[187]

	Value			
	First generation		Second generation	
Character	$\overline{X}$	Range	$\overline{X}$	Range
Total length (mm)	5.6	4.0—7.9	2.1	1.7—2.33
Greatest width	236.0	180.0—300.0	94.0	80.04—100.0
Length stoma	5.0	3.4—6.2	3.5	3.1—6.2
Width stoma	9.3	7.0—12.4	7.4	6.5—9.3
Length head to base of esophagus	244.0	220.0—284.0	214.0	205.0—220.0
Length head to excretory pore	165.0	142.0—180.0	131.0	121.0—149.0
Length head to nerve ring	177.0	148.0—192.0	159.0	149.0—167.0
% vulva	51.0	48.0—53.0	54.0	51.0—56.0
Length tail	45.0	43.0—53.0	58.0	56.0—62.0
Width at anus	75.0	56.0—93.0	38.0	34.0—43.0
Protrusion of vulva	31.0	22.0—37.0	6.5	4.6—9.3
Length tail knob	18.0	15.0—22.0	—	— —

TABLE 86

Comparative Measurements of First and Second Generation Males of *Neoaplectana glaseri*[187]

	Value			
	First generation		Second generation	
Character	$\overline{X}$	Range	$\overline{X}$	Range
Total length (mm)	1.7	1.5—1.9	1.4	1.2—1.5
Greatest width	72.0	54.0—92.0	55.0	54—62.0
Length stoma	1.3	1.2—1.4	5.5	5.0—7.0
Width stoma	1.3	1.2—1.4	4.0	3.0—5.0
Length head to base of esophagus	160.0	155.0—187.0	169.0	155.0—180.0
Length head to excretory pore	145.0	121.0—178.0	115.0	105.0—124.0
Length head to nerve ring	132.0	99.0—183.0	139.0	130.0—146.0
Length reflexion of testis	176.0	84.0—264.0	115.0	93.0—139.0
Length tail	30.0	28.0—44.0	28.0	25.0—31.0
Width at anus	42.0	34.0—47.0	37.0	34.0—43.0
Length spicule	77.0	62.0—90.0	61.0	59.0—62.0
Width spicule	9.0	6.0—12.0	7.2	6.2—9.3
Length gubernaculum	46.0	40.0—50.0	36.0	34.0—37.0
Width gubernaculum	8.0	6.0—9.0	6.0	6.0—6.0

intestinal tract contains cells of a symbiotic bacterium. The infective stage juveniles of *N. glaseri* are relatively large in comparison with other neoaplectanid nematodes (Figures 29 and 30). Quantitative values of this stage are presented in Table 87.

The above description is based on a population of *N. glaseri* described from *Strigoderma arboricola* (Fab.) in North Carolina and agrees essentially in all qualitative and most quantitative characters with the the original description of this nematode by Steiner[195] and the emended description of Turco et al.[179] The diagnostic characters of this nematode species include the rounded tail in the males and first generation females, the notch in the tip of the spicules, and the size of the infective stage juveniles. Unfortunately, neither Steiner[195] nor Turco et al.[179] stated whether their material comprised first or second generation populations from the insect or from artificially cultured

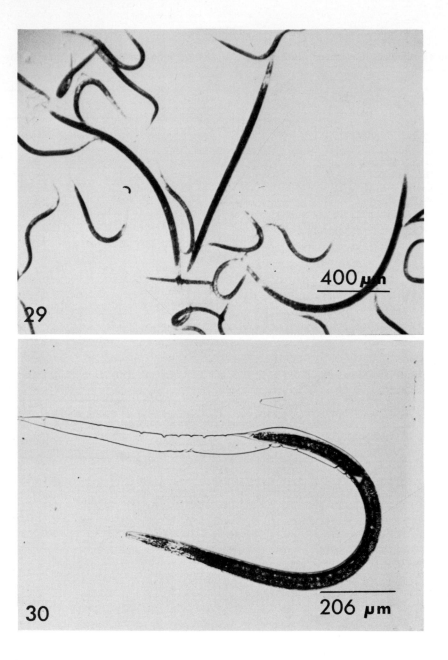

FIGURE 29 and 30.   (29) Three infective juveniles of *Neoaplectana glaseri* surrounded by the smaller infective juveniles of *Neoaplactana carpocapsae.* (30) Infective juvenile *Neoaplectana glaseri* shedding the second stage cuticle.

material; thus, the quantitative measurements are difficult to interpret. The importance of stating the source or generation of adults being studied is obvious from the current investigation. Apart from the length and width, other measurements vary greatly between first and second generation adults (see Tables 85 and 86). Perhaps the most striking qualitative difference is the shape of the female tail. The tail of first generation females possesses a small rounded projection, while that of the second generation is conically pointed with a sharp mucron and is significantly longer, surprisingly, than that of the first generation. Viewed separately, it would not be difficult to

TABLE 87

Measurements of the Infective Stage Juveniles of *Neoaplectana glaseri*[187]

	Value	
Character (N = 10)	X̄	Range
Total length (mm)	1060	864—1448
Greatest width	45	35— 50
Length head to excretory pore	100	87— 108
Length of tail	76	62— 87

find significant differences and erroneously make these separate species. The first and second generation males do not differ much from each other; however, the testis is definitely reflexed, as it is in other species of *Neoaplectana*, and not extended as stated by Steiner[195] and Turco et al.[179]

2. Bionomics

Life cycle — Developmental studies indicated that *N. glaseri* has a life cycle comparable to that of *N. carpocapsae*, namely, the infective stage juveniles enter the host insect by way of the mouth or other natural openings, then enter the body cavity where they undergo further development to first generation adult males and females. These adults mate and a portion of the progeny develop into second generation adults. The progeny of these adults mature to infective stage juveniles that leave the host. Compared to *N. carpocapsae, N. glaseri* develops more rapidly in larvae of *Galleria mellonella*. Using the standard *Neoaplectana* infection method of placing both hosts and nematodes on wet filter paper, it took only 3 days (at 22°C) for the infective stages of *N. glaseri* to enter the host's hemocoel, develop into first generation adults, mate, and oviposit (in contrast to 5 days with *N. carpocapsae*). Second generation adults appeared 2 to 3 days later, and the infective stages began emerging the following day, just 1 week after initial exposure (in contrast to 10 days with *N. carpocapsae*). At 25°C, the entire cycle was completed in just 4 days.

One aspect of the biology that should be discussed is a statement made by Glaser[196] that "after the invasive forms enter a grub, they develop into males and females, and the females shed their young into the alimentary system." This statement implies that there is at least one generation that occurs in the intestinal tract of the host. This is a very unusual location for developing neoaplectanid nematodes and is probably a faulty observation. In our own studies, the infective stage juveniles always entered the hemocoel before initiating development, as do all the other neoaplectanid nematodes studied.

Ecology: At the time of its original discovery, in larvae of the Japanese beetle (*Popillia japonica* Newm.) at the Tavistock Country Club in Haddonfield, New Jersey, it was not known whether *N. glaseri* was native to the Orient or to North America. However, it was subsequently discovered parasitizing native insects in North America (Table 88) and was probably established in North America long before the Japanese beetle arrived. Also, extensive searches were made in Japan and the Orient for natural enemies of the beetle over a period of 13 years. During this time, over 1½ million predaceous and parasitic insects were collected; however, no mention was made of discovery of *N. glaseri* or, for that matter, any other species of *Neoaplectana*.[197,198]

After a considerable amount of research and field testing of *N. glaseri*, interest slackened and soon all natural populations of the nematode were gone. In New Jersey, where most of the work on this species was performed, *N. glaseri* was last recovered

TABLE 88

Host Range of *Neoaplectana glaseri*

Order	Family	Host	Stage	Infection[a]	Ref.
Coleoptera	Chrysomeli- dae	*Diabrotica undecempunctata howardi* (Barber)	A	E	179
	Curculioni- dae	*Curculio* sp.	L	E	202
		Cylas formicarius Fab.	L	E	202
		Lissorhoptrus oryzophilus (Kuschel)	A	E	179
		Listroderes obliquus Klug.	L	E	202
		Panotmorus leucoloma Boh.	L	E	203
		P. peregrinus Buch.	L, A, P	E, N	200
		P. striatus Buch.	L, A, P	E	202
		P. taeniatulus (Berg.)	L, A, P	E	202
	Elateridae	Elaterid beetles	P	N	203
	Scarabaeidae	*Anomala orientalis* Waterh.	L	E	203
		Autoserica castanea Arrow	L	E	203
		Costelytra zealandica (White)	L	E	206
		Cotalpa sp.	L	E	203
		Cotinis nitida L.	L	E	203
		Cyclocephala borealis Arrow	L	E	202
		Macrodactylus subspinosus Fab.	L	E	203
		Melolontha spp.	L	E	206
		Ochrosidia villosa Burm	L	E	203
		Odontria sp.	L	E	207
		O. zealandica W., V., E., and T.	L	E	207
		Oxycanus cervinatus Walk.	L	E	206
		Phyllophaga spp.	L	E	203
		Pleurophorus sp.	L	E	202
		Popillia japonica Newm.	L	N	196
		Pyronota festiva (Fab.)	L	E	207
		Strigoderma arboricola (Fab.)	L	N	186
		Xyloryctes satyrus Fab.	L	E	203
Lepidoptera	Galleriidae	*Galleria mellonella* (L.)	L	E	186
	Noctuidae	Noctuid larvae	L	N	203
		Pyrausta nubilalis Hbn.	L	E	203
		Spodoptera frugiperda (Abl. and Smith)	L	E	179
	Pyralidae	*Diatraea saccharalis* (Fab.)	L	E	179

[a] E, experimental infection; N, natural infection.

from the field in 1953 from a locality that had been infested in 1940.[199] A nematode collected from white-fringed beetle larvae (*Pantomorus* spp.) in Mississippi in 1941 was called *Neoaplectana* 41035[200] and later was shown to be conspecific with *N. glaseri*.[201] However, cultures of this population apparently were lost. *N. glaseri* was recovered from a field in Louisiana in 1969,[200a] but cultures apparently were not permanently established.

In 1967, the present author visited localities in New Jersey where *N. glaseri* had been released in the 1930s, but the few Japanese beetle larvae collected were free from nematode infection.

Some time ago, nematodes determined by the author to be *N. glaseri* were found infecting larvae of the rose chafer *Strigoderma arboricola* in North Carolina by W. M. Brooks at Raleigh. Since the original description of *N. glaseri*[195] and subsequent descriptions[179] failed to indicate quantitative and qualitative differences found between the first and second generation adults, an amended description was published.[187]

N. glaseri is obviously a nematode suited for the parasitism of soil insects, although some above-ground insects are also attacked (see Table 88). Distribution is mainly through the soil; however, Glaser[196] noted that adult Japanese beetles contained *N. glaseri* in their alimentary tracts and concluded that the nematodes are possibly spread by the adult beetles. Swain et al.[202] speculated that the texture of the soil influences the rate of travel of the infective stage nematodes and found that nematodes applied to test plots moved from 10 to 20 ft in a few months to invade fumigated check plots. Nematode movement in wooden 14-ft² bins during the winter months in Mississippi reached 6 ft in 33 days. The authors noted that, rarely, adult white-fringed beetles that emerged from infested soil in screen cages died with nematodes 1 to 3 days later.

Vertical distribution of *N. glaseri* following surface application in fumigated outdoor plots during the summer showed soil penetration to a depth of 20 in.[202] Additional studies showed that while *N. glaseri* could be found from June to December in soil from 0 to 18 in. deep, the majority were found in the 6- to 12-in. zone.

Although soil pH was not listed as an important ecological factor for *N. glaseri*, temperature was. The most abundant parasitism of field populations of white-fringed beetles occurred when the soil temperature averaged 85°F 3 in. below the surface. Lower temperatures inhibited nematode development. Soil moisture was considered an important factor, and the heaviest nematode populations occurred where the water table was within a few feet of the soil surface during much of the year.[262] According to Glaser et al.,[203] the infective stages of *N. glaseri* can survive for 1½ years in the field without hosts.

N. glaseri is probably distributed throughout the middle and southeastern states where it attacks soil insects, especially Coleoptera, but also Lepidoptera and probably representatives of other orders. Natural populations of this nematode attacked larvae of the white-fringed beetle, and studies by Swain et al.[202] showed that *N. glaseri* was generally distributed and played a major role in reducing beetle populations.

3. Bacterial Associates

In an unpublished Ph.D. thesis, Dutky[204] mentioned finding a specific bacterium and nematodes together in the body cavity of diseased Japanese beetle grubs and indicated that the former may be carried into the beetle by the latter. However, further data indicated that the above nematodes probably were not *N. glaseri* but another species whose infective stages were about half the size of those of *N. glaseri*. This other species was probably what was then called *N. chresima* (now considered a strain of *N. carpocapsae*). Both *N. chresima* and *N. glaseri* had been found in field populations of Japanese beetles.

Thus, it was not established at that time whether *N. glaseri* was associated with a specific bacterium that played an important role in host pathogenicity and parasite nutrition. The fact that after being killed by *N. glaseri*, Japanese beetle larvae turned a uniform rusty or ocherous brown color is in itself suggestive of the presence of an associated bacterium (Figure 31). Thus, it is surprising that no mention of the specific bacterium occurred in the many papers on the biology and culture of this nematode. However, in a recently found population of *N. glaseri*, the presence of a specific bacterium was confirmed. These bacteria will grow on nutrient agar slants where they produce pleimorphic cells, often containing refractile deposits interpreted as stored glycogen. Using the hanging blood drop method of isolating bacteria inside infective stage neoaplectanids,[224] the same bacterium was consistently recovered from the intestine of infective stage *N. glaseri* (Figure 32).

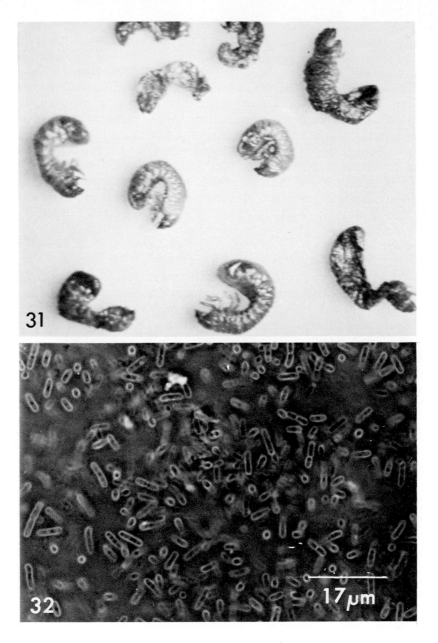

FIGURE 31 and 32.   (31) Japanese beetle grubs infected with *Neoaplectana glaseri*.
(32) Cells of the symbiotic bacterium associated with *Neoaplectana glaseri*.

This bacterium is a nonspore-forming, peritrichously flagellated, Gram-negative rod that forms rusty brown colonies on nutrient agar. Experiments showed that this bacterium could be taken up by original cultures of *N. glaseri* that had been maintained axenically for many years. After injecting axenic *N. glaseri* and bacterial cells into healthy insect larvae, the nematodes developed, and the emerging infective stages of the previously axenic nematodes contained cells of the bacterium associated with xenic *N. glaseri*.

The bacterium associated with *N. glaseri* probably functions in a manner similar to *Achromobacter nematophilus* and *N. carpocapsae*,[205] namely to establish suitable conditions for development inside the insect by supplying nourishment and inhibiting other bacteria. However, we have discovered that axenic populations of *N. glaseri* can reproduce to a limited extent inside the hemocoel of living wax moth larvae but that reproduction is much more extensive with the presence of the associated bacterium.

Attempts were made by the present author to introduce cells of *A. nematophilus* (associated with *N. carpocapsae*) into axenic populations of *N. glaseri*. Although some of the nematodes reproduced, none of the developing infective stages contained cells of *A. nematophilus* in their intestines. Tests were also conducted to determine if *N. glaseri* could accept *Alcaligines faecalis, Proteus rittgeri, Pseudomonas aeruginosa* and *Serratia marcescens*. There was some reproduction in the presence of the former three species, but the cells were not retained by the infective stage juveniles.

4. Host Range

Although *N. glaseri* was originally described from a scarabaeid larva and is known to infect a number of species in this family in experimental situations, it can also develop in other beetles as well as in various moth larvae (see Table 88). Finding pupae of elaterid beetles infected with this species suggests that the infective stages may enter natural openings of the insect other than the mouth. Adult Japanese beetles and other insects have also been infected.

5. Culture

N. glaseri was cultured on artificial media soon after its discovery, and most of the field applications of this nematode were made with material cultured on artificial media. The initial basic medium was veal infusion agar in 5½-cm petri dishes[196] (Figure 33). One day before inoculation, 2 cc of a 10% dextrose solution was added to the petri dish, together with about 8 cc of remelted veal infusion agar. When cool, the surface of the medium was flooded with a heavy water suspension of a living, pure culture of baker's yeast (Fleischmann's yeast purified from bacteria by plating on dextrose agar). The plate was incubated at room temperature until the yeast had grown uniformly over the entire surface. A Japanese beetle grub containing *N. glaseri* was washed, triturated, and some of the nematodes transferred with a pipette to the yeast-culture plate. The plate was incubated at 24 to 27°C, and a few drops of sterile tap water were added every 2 or 3 days. The yeast inhibited bacterial development which was considered detrimental to the nematodes. Glaser[196] stated that *Neoaplectana* cultures were never obtained on living or dead bacteria, nor would they develop on heavy suspensions of killed yeast cells. The living yeast furnished some important nutritional substitute found in the insect host.

Although transfers were made about every 2 weeks, during which 3 or 4 generations were produced, most of the strains "died out" after the seventh or eighth transfer or in 14 to 16 weeks. The females appeared normal in size and shape, but the ovaries failed to mature and eggs were no longer visible. Glaser found that artificial cultures could be reestablished by infecting Japanese beetle grubs with infective stages removed from the sixth transfer. It was discovered that the nematodes survived longer on artificial media if allowed to pass through several hosts in succession rather than a single beetle larva. A later study[208] indicated that some growth factor initially stored within the tissues of the developing nematodes was gradually depleted with each generation. This growth factor could be restored to the nematodes by providing to the standard artificial media supplements of living yeast, host substance, or bovine, ovarian substance.

Initial success in field experiments prompted the need for larger numbers of *N. glaseri* and a more practical and economic method of cultivation. The above method of veal agar cultivation was too expensive and did not yield large quantities of nematodes needed for field distribution. Thus, McCoy and Glaser[209] devised a fermented potato mash for fulfilling this purpose. The technique involved chopping up, by weight, two

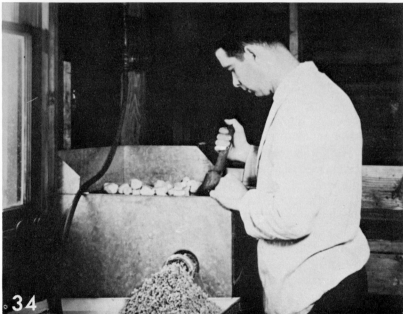

FIGURE 33 and 34.   (33) Glaser and co-workers preparing veal infusion agar for the culture of *Neoaplectana glaseri*. (34) Preparing potato mash for the culture of *Neoaplectana glaseri*.

thirds Irish potatoes and one third sweet potatoes (Figure 34). The latter enhanced growth of the yeast. The media was mixed with a pure culture of baker's yeast grown on sterilized sweet potato gruel and spread in a 2-cm deep layer in clean, covered culture trays (33 cm × 61 cm × 3.8 cm). It was allowed to ferment at 70°F for 20 to 40 hr and then mixed with a preservative (sodium derivative of the methyl ester of par-ahydroxy-benzoic acid). After adding the preservative, the pH of the medium was adjusted to 7.8 and then spread out on a clean culture tray. Washed nematodes (200,000 to 400,000) from agar cultures were added to the entire surface of the medium and the cultures were incubated at 21 to 24°C for a period of 6 to 10 days. When the cultures contained a high percentage of infective stages, the trays were released for immediate field distribution (Figure 35).

FIGURE 35 and 36.   (35) Inoculating potato mash medium with *Neoaplectana glaseri*.
(36) Examining Japanese beetle grubs after a field application of *Neoaplectana glaseri*.

The production rate using the potato mash method averaged about four million nematodes (approximately 2000 cm² of culture surface). The production from one or two 200-mm agar petri plates varied from 200,000 to 400,000 nematodes.

A third method of culturing *N. glaseri*, using ground, extracted veal pulp, was described by McCoy and Girth.[210] This method consisted of grinding fresh veal through a food chopper, then infusing it for up to 48 hr with twice its weight of distilled water. The infusion was poured into a flannel cloth and drained, and the pulp was squeezed as dry as possible. The expressed juice was suitable for making veal infusion agar and, after adding water and a preservative (0.06% formaldehyde and 0.05% of the sodium derivative of methyl-*p*-hydroxybenzoate), the medium was spread into a thin, uniform layer and inoculated with nematodes from other sources.

The authors stated that all stages of *N. glaseri* grown on veal medium were larger and more robust than those grown by the original agar method. This method produced a yield of from 9000 to 12,000 nematodes per square centimeter of culture area. This made the average yield per tray (2000 cm^2) approximately 20 million infective stage juveniles.

A summary of all three above methods, with some modifications, was presented by Glaser et al.[203] In this paper, the authors gave methods for removing the infective stage juveniles developing on agar plates. Sterile water was first passed over the culture surface, and the nematodes and debris were worked into a suspension which was poured into a large test tube. The nematodes were allowed to settle to the bottom and the supernatant fluid discarded. After repeating the above process several times, the nematodes were allowed to migrate through lens paper supported on a sieve, with the bottom in contact with water. The nematodes wiggled through the lens paper leaving sediment, etc. on the surface of the paper.

The authors continued by discussing the problem of obtaining ensheathed nematodes from the cultures. "Ensheathed nematode" means the third stage infective juveniles that are still surrounded by the second stage cuticle. The nematode should be in this physiological state before being introduced into the environment. If not, the survival rate will be very low. Glaser et al.[203] simply kept the nematodes removed from culture in an isotonic salt solution until they ensheathed. They mentioned that the nematodes should not be too crowded and the temperature must be high enough to permit ensheathment (the optimum was around 24°C). They also noted that just prior to ensheathment, the second stage juveniles voided the contents of their alimentary tracts and fouled the water. Ensheathment was completed within 4 days at 24°C. If large numbers of nematodes were present, it was helpful to bubble compressed air through the liquid. The final ensheathed nematodes were kept in bottles of water for up to 6 months at room temperature.

The above studies laid the foundation for the axenic culture of *N. glaseri*. Glaser[208] noted that when the ensheathed infective stages were carefully washed and repeatedly treated with sodium hypochlorite for 30 to 60 min, they were sterilized. When these sterile juveniles were placed on veal infusion agar slants containing about 1 g of animal tissue removed under sterile conditions, they initiated development and after 18 days produced approximately 150,000 nematodes. Although 20-day-old mouse embryo, beef kidney, and rabbit ovary and kidney all supported nematode growth, rabbit kidney was the easiest to manipulate and gave the best growth. Although this and subsequent methods of axenic culture[211] represented a scientific breakthrough for the study of nutritional and other requirements of metazoan animals, these methods were costly and impractical from the standpoint of biological control.

Using the veal infusion agar medium inoculated with yeast and ovarian substance, Swain et al.[202] produced *N. glaseri* at a cost of approximately $1.00 per million nematodes. They noted that the infective stages would survive for more than a year when refrigerated at 40°F; however, a disease frequently occurred in stored nematodes. This disease produced misshapen nematodes that eventually died.

6. Application

N. glaseri had an illustrious history in the 1930s and early 1940s as it was applied over large areas in attempts to control the Japanese beetle and other insects. The first field applications of *N. glaseri* were made in two localities in southern New Jersey where the Japanese beetle grubs averaged approximately 80/ft^2.[196] Nematode cultures grown on veal infusion agar were flooded with water and washed into a sprinkling can containing 2 gal of water. Approximately 160,000 nematodes were applied to each of the two 6-ft² plots. The nematodes became established and produced high mortality. Further examination of these plots by Glaser and Farrell[212] supported Glaser's earlier findings and demonstrated the presence of infective stage juveniles 3 years after initial application (Figure 36). A second field trial[212] was conducted in 1931 on two 15-ft² plots of lawn grass. The insect population covered a 22-ft² area (to a depth of 6 in.). After washing the nematodes off veal infusion agar plates and suspending them in 8 gal of water, they were sprinkled with a watering can over the test plot. None of the nematodes were ever recovered, and no cases of parasitism were discovered.

A second field application was made on timothy and clover pasture plots (two at 130 × 30 ft) with an average grub count of 31/ft². One of the test plots received a surface application of nematodes in a watering can. The other was treated with nematodes placed in a power sprayer. The tank contained 250 gal of water, and the spray nozzle did not damage the nematodes. The results of this experiment were poor, and the nematodes did not become established.

A third experiment consisted of placing nematodes in 3- to 4-in. deep holes dug in a pasture. The 75 holes in the 110 × 20 ft plots were spaced approximately 5 ft apart, and each hole received a half pie plate culture of nematodes. The authors concluded that the nematodes had significantly reduced the grub populations by about 40% and that the subsurface method of introduction was more effective than surface spraying.

During the fall of 1936 and the spring of 1937, the New Jersey Department of Agriculture distributed about 3.6 billion nematodes over various parts of the state. All of these nematodes were reared on the veal pulp method.[210] The nematodes were applied at 3½-mi intervals over the state in order to increase the distribution of N. glaseri.[203]

In summarizing the field experiments with N. glaseri, Glaser et al.[203] stated that of 73 different experimental plots treated at various seasons since 1931, parasitized Japanese beetle grubs were recovered from 72. Depending on soil moisture, nematode dosage, soil temperature, and host density, from 0.3 to 81% of the total beetle population in these plots were parasitized. The authors stated that best results were achieved when ensheathed juveniles were used rather than the complete nematode population from the culture and that the surface method of application was better than the subsurface method. In surface application, the nematodes are applied from a watering can or a tank sprayer directly onto the vegetation. In one area that had a heavy initial dose of nematodes, N. glaseri became established and survived under field conditions for 8½ years.

In Mississippi, N. glaseri was also used against larvae of the white-fringed beetle (Pantomorus spp.).[202] The nematodes had been applied in strips 30 ft apart for large areas (over 1000 ft²) or uniformly over the test area (complete cover method). In both methods, the nematodes were washed into the soil with a power sprayer.

The nematodes were shipped in 12-oz glass medicine bottles with tops tightly capped. Each bottle contained 50 to 60 ml of nematode suspension (10,000 nematodes per milliliter of water). Losses only occurred when the temperature was very high or when there were long waits en route.

Field introductions of N. glaseri were made in Alabama, Florida, Louisiana, Mississippi, and North Carolina. In these states, 31 plots of 1000 ft² or more were established at 21 locations. These plots were treated with N. glaseri at the rate of 25,000 nematodes per square foot. Using the Baermann funnel method, nematodes were recovered from all of the plots except two. Cultures of N. glaseri were sent to New Zealand to be used against the grass grub Costelytra zealandica. The nematodes were cultured on veal infusion agar and ground veal pulp and applied to grub-infected plots with a watering can at the rate of 80,000 nematodes per square foot. Although nematode-infected grass grubs were recovered up to a year after the experiment, there was no general reduction in population as compared with the controls.

Although the rate of parasitism of the host insects attacked by *N. glaseri* was not always high enough for adequate control, it was remarkable that so many populations of the nematode became established in nature. This is because all nematodes used in field trials were cultured on artificial media supplied with chemicals to inhibit fungi and bacteria. Exactly what effect these rearing methods had on the associated bacterium is not known, but it is probable that all the initial associated bacteria, including the symbiotic one, were soon eliminated and replaced with chance contaminants. Without the symbiotic bacteria, reproduction and development in host insects would be greatly reduced and these nematodes would find it difficult to survive in nature.

A summary of the field trials using *N. glaseri* is presented in Table 89.

F. *Neoaplectana menozzii* Travassos 1932

The second described species in the genus *Neoaplectana, N. menozzii*, was recovered by Travassos[214,215] from parasitized sugar beet weevils, *Conorrhynchus* (syn. *Clenous*) *mendicus* Gyll. in Italy. Apparently, it has never been recovered again.

1. Description[214,215]

Adults — A relatively small *Neoaplectana* with posterior extremity blunt and conical. The excretory pore is anterior to the nerve ring. The pharynx has an anterior swollen cylindrical portion, a narrow isthmus portion, and an enlarged terminal bulb containing rudimentary valves. The male tail has eight to ten pairs of preanal and an undetermined number of postanal papillae. Spicules have a dilated proximal portion. The gubernaculum is fusiform. Females contain eggs and developing juveniles. The vulva is posterior in position. Eggs are 32×26 μm. Quantitative values of the adults are presented in Tables 90 and 91.

2. Bionomics

N. menozzii was found in the larval, pupal, and adult stages of *Conorrhynchus mendicus* Gyll. No details are presented concerning its life cycle or ecology, and apparently it was never used as a biological control agent.

G. *Neoaplectana feltiae* Filipjev 1934

This is the third species described in the genus *Neoaplectana* and it was recovered from the armyworm *Agrotis segetum* (Schif.) in eastern Russia.[181] Unfortunately, Filipjev's description of this species is not as complete as one would desire, yet several characters are mentioned which can serve as diagnostic features. Recently, Stanuszek[183,184,216,217] redescribed what he considered to be *N. feltiae* and showed how this species could interbreed with *N. carpocapsae*. However, as is pointed out later, Stanuszek's *N. feltiae* does not agree with several basic characters shown in Filipjev's earlier description and may indeed be another geographical population of *N. carpocapsae*.

A synopsis of the species based on Filipjev's description is presented below. Quantitative measurements are presented in Tables 92 and 93. Unfortunately, Filipjev did not state whether he was measuring first or second generation adults; however, his description of the infection suggests that he encountered second generation adults.

1. Description[181]

Adults — The female tail is short and conical with a small mucron. The cuticle is smooth. The head is not offset, with a ring of six papillae (second ring not observed). The stoma is shallow and chitinized. The anterior portion of the pharynx is cylindrical,

TABLE 89

Field Trials of *Neoaplectana glaseri* Against Insect Populations

Host	Site	Dosage	Rearing method	Results	Location	Ref.
Japanese beetle *Popillia japonica* (Newm).	Pasture turf, two test plots each 6 ft²	Approximately 27,000 nemas per ft² soil	Artificial media (xenic)	Nemas established high mortality	Eastern U.S. (New Jersey)	196
Popillia japonica	Lawn grass, pasture turf	On soil surface 3 to 4 in. in soil	Artificial media (xenic)	No control; significantly reduced grub populations by about 40%	Eastern U.S. (New Jersey)	212
White fringed beetles *Pantomorus* spp.	Soil, 31 plots each 1000 ft²	25,000 nemas per ft² soil surface	Artificial media (xenic)	Nemas established in 29 plots	Southern U.S.	200, 201
Grass grub *Costelytra zealandica* (White)	Grass plot, 2500 ft²	80,000 nemas per ft² soil surface	Artificial media (xenic)	Infected grubs recovered no reduction over controls	New Zealand	213
Rhinoceros beetle *Oryctes rhinoceros* L.	Decaying matter	—	—	—	Wallis Island	213
White grubs *Phyllophaga* spp.	Turf	—	—	—	Midwest U.S.	187
Adoretus sp.	—	—	—	—	Hawaii	187

TABLE 90	
Measurements of *Neoaplectana menozzii* Females[215]	
Character	Value
Total length (mm)	1.1—1.6
Greatest width	80—86
Length stoma	4
Length head to base of esophagus	153—159
Length head to excretory pore	70—80
Length head to nerve ring	110—120
Length tail	110—120
% vulva	58—60

TABLE 91	
Measurements of *Neoaplectana menozzii* Males[215]	
Character	Value
Total length (mm)	0.80—0.86
Greatest width	54—72
Length head to excretory pore	72—75
Length head to nerve ring	90—100
Length tail	18—21
Length spicule	54—56
Length gubernaculum	40

followed by a swelling, then a contraction, and finally a prominent terminal bulb with a valve. The excretory pore opening is posterior to the nerve ring. Ovaries are paired. Uteri are filled with eggs and developing young. The male tail is short and conical, terminating in a fine mucron. The head has inner and outer rings of papillae. The testis is single and reflexed. Spicules are bent, and the tail has one median, five or six paired lateral preanal genital papillae, two paired adanal lateral, one paired lateral dorsal, and six paired postanal papillae.

Infective stage juveniles — This stage is 750 to 850 μm long. The diagnostic characters for *N. feltiae* are the posterior position of the excretory pore and the presence of a long mucron on the tip of the male tail. The number and position of the male genital papillae are also unique and, if correctly represented by Filipjev, would be diagnostic.

2. Bionomics

Filipjev[181] mentions finding about 100 to 200 females, 30 males, and thousands of juveniles in a parasitized larva of *A. segetum*. Apparently, the nematodes were not cultured, so that no detailed studies were conducted on their biology. The diseased caterpillars were collected in 1926, and the infection was very localized. No parasitized insects were collected from the same site the following year. There is no mention of an associated bacterium, although Filipjev mentions that the infected caterpillars had a shiny appearance, which may have been due to an associated bacterium.

3. Other records of Neoaplectana feltiae

Stanuszek[217] mentioned finding *N. feltiae* from caterpillars near Warsaw and later described this nematode and another he considered a new subspecies of *N. feltiae*.[183,184] However, there was no mention of Filipjev's earlier description, and Stanuszek did not point out that differences between his and Filipjev's *N. feltiae* occur. It is now known that most quantitative measurements cannot be used to compare adults of the genus *Neoaplectana*. However, the character, head to excretory pore, tends to remain fairly constant in *Neoaplectana* spp., and there is no overlap between Filipjev and Stanuszek in this character. Also, the mucron on the male tail is shown by Filipjev to be about 12 to 13 μm whereas Stanuszek[183] describes.it as only reaching 4 μm. Previous studies have shown the tail mucron to be a relatively constant character. Therefore, the present author concludes that the *feltiae* of Stanuszek is probably a strain of *N. carpocapsae* Weiser. However, the quantitative values of Stanuszek's *feltiae* are presented in Tables 94 and 95.

H. Neoaplectana bibionis Bovien and N. affinis Bovien 1937

While working with larvae of bibionid flies, Bovien[218] encountered nematodes of the genus *Neoaplectana* and described them as *N. bibionis* and *N. affinis*.

TABLE 92

Measurements of *Neoaplectana feltiae* Females[181]

	Value	
Character	$\overline{X}$	Range
Total length (mm)	4.47	3.8—5.94
Greatest width	277	
Length stoma	10	
Width stoma	20	
Length head to base of esophagus	229	
Length head to excretory pore	180	
Length head to nerve ring	159	
% vulva	59	53—64
Length tail	60	
Width at anus	56	

TABLE 93

Measurements of *Neoaplectana feltiae* Males[8x8]

	Value	
Character	$\overline{X}$	Range
Total length (mm)	1.37	1.33—1.42
Greatest width	113	
Length head to base of esophagus	147	
Length head to excretory pore	120	
Length head to nerve ring	105	
Length tail	35	
Width at anus	36	
Length spicule	160—6	
Length gubernaculum	352—5	
Length mucron	13	

1. Description[218]

Adult N. bibionis — Females may be elongate or shortened (dwarfs). The tail varies from bluntly conical or slightly rounded with a mucron in mature forms to more pointed in young females. The excretory pore is anterior to the nerve ring. The vulvar opening is a transverse slit situated somewhat behind the middle of the body. The uterus is filled with eggs (diameter, 42 to 46 μm) and developing juveniles. Ovaries are reflexed. In males the cuticle is smooth. The testis is reflexed. The excretory pore is anterior to the nerve ring. Spicules are yellow or brownish in color. The gubernaculum is yellow and broad in the middle and narrow at the ends. The male has 11 pairs and one preanal ventral papilla. The tail may or may not have a mucron.

Adult N. affinis — Females are similar to *N. bibionis*. Males are similar to *N. bibionis* except that the spicules and gubernaculum are pale grey in color, somewhat more curved, and the proximal portion is much smaller than those of *N. bibionis*. The gubernaculum is more evenly curved and is more crescent shaped than that of *N. bibionis;* the proximal knob or hook is lacking. Quantitative values of the adults are presented in Tables 96 and 97.

Infective stage juveniles — This stage of *N. bibionis* is 700 to 1000 μm long, 23 to 26 μm wide (somewhat larger than *N. affinis*), and also has a longer tail than the latter species. *N. affinis* possess a small spinelike structure of refractive material in the tip of their tail (Figure 39).

TABLE 94

Measurements of *Neoaplectana feltiae* Females[183,184]

Character	First generation		Second generation	
	X̄	Range	X̄	Range
Total length (mm)	11.3	9.0—18.0	2.30	0.8—3.5
Greatest width	342	190—450	149	75—180
Length stoma	10	6.5—11.3	7.7	4.0—10.1
Width stoma	12.6	9.8—14.5	7.9	5.2—9.5
Length head to base of esophagus	310	225—388	83	55—116
Length head to excretory pore	124	92—170	86	47—106
Length head to nerve ring	248	212—272	169	132—229
% vulva	51	50—53	55	48—60
Length tail	77	53—100	40	30—49
Width at anus	190	115—258	65	40—89
Length tail knob	11.3	7.9—18.2	1.5	1.1—6.6

TABLE 95

Measurements of *Neoaplectana feltiae* Males[183,184]

Character	First generation		Second generation	
	X̄	Range	X̄	Range
Total length (mm)	1.79	1.6—2.8	1.08	0.68—1.40
Greatest width	136	105—220	49	35—51
Length stoma	5.8	4.8—6.5	4.5	1.8—7.3
Width stoma	3.9	3.6—5.9	6.0	2.2—7.8
Length head to base of esophagus	199	171—205	125	120—146
Length head to excretory pore	79	60—101	49	36—74
Length head to nerve ring	173	131—197	107	87—116
Length reflexion of testis	750	428—985	530	240—712
Length tail	40	33—49	32	28—41
Width at anus	50	47—56	46	33—59
Length spicule	82	70—97	67	51—71
Length gubernaculum	50	46—59	43	38—56
Length mucron	2.9	2.0—3.8	3.0	1.7—4.0

2. Description[185]*

Adults — The mouth opening is surrounded by six partially fused lips, each bearing one conspicuous papilla and a second less conspicuous papilla. The pharynx has a slight medial swelling and a basal bulb containing a cavity lined with refractive ridges. Females are amphidelphic with opposed reflexed ovaries. The vulva is a transverse slit and may be protruding from the body surface (especially in first generation females). Eggs may hatch and develop inside the female uterus. The excretory pore is opposite the median portion of pharynx. The tail is variable in shape, sometimes with a postanal swelling. A small spine may or may not be present. Males have a single reflexed testis. Spicules are variable in shape, with two lateral ribs that often extend into the manubrium or head of the spicule. Spicule tips are blunt. The gubernaculum is slightly curved and boat shaped. The tail bears 22 + 1 genital papillae, two pairs of which

* Based on material collected by Poinar and Lindhardt[185] and cultured in larvae of *Galleria mellonella*.

TABLE 96

Measurements of *Neoaplectana bibionis* Females[218]

Character	Value
Total length (mm)	1.0—5.0
Greatest width	Up to 220
Length head to base of esophagus	208
Length head to excretory pore	42
Length head to nerve ring	187
% vulva	57
Length tail	21
Width at anus	83

TABLE 97

Measurements of *Neoaplectana bibionis* and *N. affinis* Males[218]

Character	Value	
	N. bibionis	*N. affinis*
Total length (mm)	0.60—1.25	1.00—1.62
Greatest width	40—100	59—100
Length head to base of esophagus	133	—
Length head to excretory pore	80	—
Length head to nerve ring	106	—
Length reflexion of testis	213	—
Length tail	28	27
Width at anus	34	36
Length spicule	53	47
Length gubernaculum	33	31

are situated on the tail tip, suggesting the presence of a cuticular spine; the latter structure is usually absent and when present, is very minute. Quantitative characters for the adults are presented in Tables 98 and 99.

Infective stage juveniles — This stage is 693 (600 to 780) μm long and the greatest width is 29 (25 to 34) μm. The distance from the head to the excretory pore is 63 (58 to 71) μm. The distance from the head to the nerve ring is 95 (84 to 99) μm. The pharynx is 130 (109 to 146) μm long. The tail is 59 (49 to 68) μm long. Some of the infective stage juveniles have a fine mucron within the tip of the ensheathing second stage cuticle, while others lack this character. The present author feels that this might depend upon the stage of development of the nematode, the spine forming as the nematodes changes into the infective stage (Figures 39 and 40).

Diagnosis — Although Bovien[218] showed a spine on the end of the male tail of *N. bibionis,* he stated that it may or may not be present. The present author rarely found a spine on males in populations of *Neoaplectana* collected from bibionid flies in Denmark and considers its presence to be very rare. In general, the measurements were greater in material collected by Poinar and Lindhardt,[185] but this may be because the specimens were reared in larvae of *Galleria mellonella,* whereas Bovien's original observations were made on material removed from larvae of bibionid flies or from artificial media. These figures again point out the heterogeneity found in populations of a neoaplectanid species.

The fact that Poinar and Lindhardt's material constitutes the same species that Bovien described is based on (1) the origin of the material which was from one of Bovien's original localities (in Skive, Denmark), (2) the nematodes came from the same host

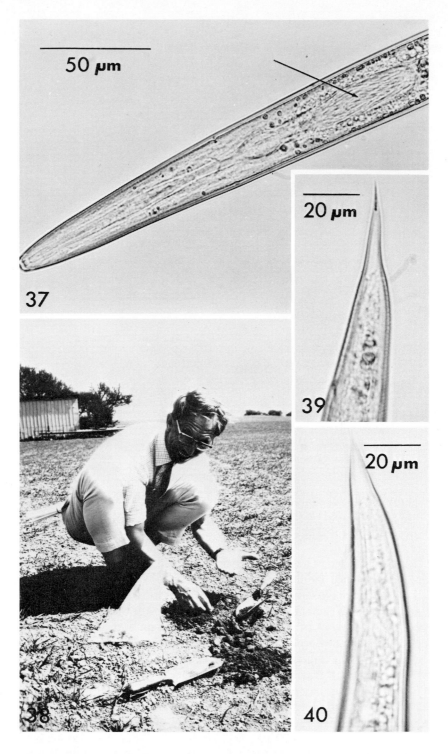

FIGURE 37 to 40.   (37) Infective juvenile of *Neoaplectana bibionis* showing cluster of symbiotic bacteria in the ventricular portion of the intestine (arrow). (38) Collecting bibionid larvae infected with *Neoaplectana bibionis* in a Danish barley field. (39) Spine-bearing tail of an infective juvenile of *Neoaplectana bibionis*. (40) Spineless tail of an infective juvenile of *Neoaplectana bibionis*.

TABLE 98

Measurements of *Neoaplectana bibionis* Females[185]

Character (N = 10)	First generation		Second generation	
	X̄	Range	X̄	Range
Total length (mm)	5.44	4.42—6.53	2.97	2.76—3.24
Greatest width	152	123—177	85	69—92
Length stoma	6	4—7	4	3—5
Width stoma	7	6—8	5	3—6
Length head to base of esophagus	196	179—214	180	161—192
Length head to excretory pore	99	90—109	97	81—112
% vulva	50	46—55	54	47—58
Length tail	54	43—59	59	47—74
Width at anus	59	50—58	42	37—47

TABLE 99

Measurements of *Neoaplectana bibionis* Males[185]

Character	First generation		Second generation	
	X̄	Range	X̄	Range
Total length (mm)	3.38	2.80—4.00	2.26	2.02—2.56
Greatest width	107	80—140	66	62—77
Length stoma	4	3—6	3	2—5
Width stoma	5	4—6	5	4—6
Length head to base of esophagus	167	158—183	160	143—164
Length head to excretory pore	100	80—118	93	89—99
Length tail	51	43—59	42	40—43
Width at anus	62	50—78	46	40—53
Length spicule	70	65—77	59	55—65
Length gubernaculum	41	34—47	33	31—37

species, (3) the infective stage juveniles have the same tail spine in the second stage cuticle, and (4) the range in spicule shape agreed with that figured by Bovien.[218] Thus, the lack of a tail spine in most males, along with the EP/pharynx ratio of 0.50 to 0.70 and the length of the infective stages, separates this species from the other members of the genus.

Because of the variability found within the populations we studied (many males possessed spicules intermediate in form between *N. bibionis* and *N. affinis*) and the observations that the presence of a tail spine in the infective stage juveniles was a maturity factor, the present author feels that *N. affinis* is probably just an intraspecific form of *N. bibionis*.

3. Bionomics

Life cycle — Bovien[218] noted that the nematodes entered the body of living *Bibio* larvae but did not develop further until the host was dead. He also found nematodes in the gut of living host larvae and noted that if the bibionid larvae were kept under favorable conditions, the nematodes remained in the intestine without further development, and the host may even undergo metamorphosis. From additional experiments, Bovien concluded that his neoaplectanids were generally saprozoic, and the infective

TABLE 100

Host Range of *Neoaplectana bibionis* Bovien

Order	Family	Host	Stage	Infection[a]	Ref.
Diptera	Bibionidae	*Bibio ferruginatus* L.	L	N	218
		B. hortulanus L.	L	N	218
		Philia febrilis (L.) Syn. *Dilophus vulgaris* (Meig.)	L	N	218
Coleoptera	Tipulidae	*Tipula paludosa*	L	E	218
		Telephorus sp.	L	E	218
Lepidoptera	Galleridae	*Galleria mellonella* L.	L	E	185

[a]   N, natural infection; E, experimental infection.

stages in the intestine only initiated development after the host died. Using egg albumen as a culture medium, Bovien found that one generation took only 4 days to complete its development. He also showed that the "dauer larva" was the infective stage which entered the host. In their study of *N. bibionis,* Poinar and Lindhardt[185] noted that with *Galleria mellonella* larvae as an experimental host, the infective stage nematodes entered the body cavity directly in a manner similar to *N. carpocapsae.*

Ecology — The distribution of *N. bibionis* is fairly widespread. Bovien[218] found *Neoplectana* in practically all samples of bibionid larvae examined from 1931 to 1935. Approximately 35 years later, Poinar and Lindhardt[185] recovered *N. bibionis* from three localities in Denmark, indicating that the bibionid populations in Denmark are probably continuously associated with these nematodes.

4. Bacterial Associates

Bovien[218] was the first to note an association between a neoaplectanid nematode and its symbiotic bacterium, although he did not at that time realize the significance of this association. He showed symbiotic bacteria in the infective stage juveniles of *N. bibionis.* He also noted (later confirmed by Poinar and Lindhardt)[185] that those host larvae attacked by *Neoaplectana* were rusty or ochrous, a color due to the associated bacterium. Poinar and Lindhardt[185] isolated symbiotic bacterium from infective stages of *N. bibionis* and found it quite similar to *Achromobacter nematophilus* (Figure 37).

5. Host Range

In nature, *N. bibionis* has only been found parasitizing bibionid fly larvae. These hosts, as well as experimental infections, are cited in Table 100. The rate of infection in the field varied from 6 to 100%,[185] and it is probable that the nematodes are important controlling agents in nature (Figure 38).

6. Culture

Aside from growing the nematodes in live bibionid larvae or dead lepidopterous hosts, Bovien[218] cultured his neoaplectanids in a small amount of egg albumen containing a piece of insect tissue. The nematodes fed on the bacteria around the insect tissue. Poinar and Lindhardt[185] showed that *N. bibionis* could be cultured in larvae of the wax moth *Galleria mellonella* or on dog food agar.[1] To the author's knowledge, *N. bibionis* has never been used as a biological control agent, although it may be a possible candidate against the bibionid pests known as "love bugs" in Florida.

I. *Neoaplectana chresima* Steiner in Glaser, McCoy, and Girth 1942

This species was found parasitizing larvae of the corn earworm (*Helicoverpa armigera*) and Japanese beetle (*Popillia japonica*).

TABLE 101

Measurements of Females of *Neoaplectana chresima* Steiner [179]

	Value	
Character	$\overline{X}$	Range
Total length (mm)	1.39	1.26—1.80
Greatest width	107	102—136
Length head to base of esophagus	158	144—187
% vulva	57	51—62
Length tail	43	33—52

TABLE 102

Measurements of Males of *Neoaplectana chresima* Steiner[179]

	Value	
Character	$\overline{X}$	Range
Total length (mm)	1 0.93	0.630—1.180
Greatest width	101	92—103
Length head to base of esophagus	130	123—144
Length tail	44	32—50
Length spicule	66	54—80
Length gubernaculum	44	40—46

1. Description

Adults — According to Glaser et al.,[219] the tail terminus is sharply set off, often mucronate. The excretory pore opening is more anteriorly located than in *N. glaseri*. The arrangement of male copulatory papillae is somewhat different from the condition in *N. glaseri*. Juveniles (probably infective stages) have four equidistant longitudinal striae on lateral fields. This description is incomplete and as such cannot be accepted. This species was redescribed by Turco et al.,[179] who examined five of Steiner's original slides. Their description follows.

According to Turco et al.,[179] the female tail terminus has a mucron. The excretory pore opens ventrad at or near the nerve ring. The anus is conspicuous without a prominent postanal lip. The tail is conoid with a slightly rounded tip. The male excretory pore opening is ventrad at or near the nerve ring. The testis is reflexed. Spicules are proximally capitate and curved outward, possessing a small prominence on each side. The gubernaculum is slightly curved posteriorly and straight anteriorly, ending in a small knob. The tail is convex-conoid with a slightly rounded tip. Quantitative values of the adults are presented in Tables 101 and 102.

Diagnosis — Turco et al.[179] state that the males of *N. chresima* are distinct by lacking a mucron on the tail terminus. Yet, in the original description by Steiner in Glaser et al.,[219] the adults are described as having their tail terminus sharply set off and often mucronate. The former authors seem a bit confused, however, since in the description of the female of *N. chresima* they mention that the body of the female terminates "in a mucronulated tail terminus." A few lines later they state that the female tail is "conoid with slightly rounded tip."

On the basis of the above descriptions, the present author feels that *N. chresima* is simply a race or strain of *N. carpocapsae*. The excretory pore opening of *N. carpocapsae* is more anteriorly located than that of *N. glaseri* and agrees with that given for *N. chresima*. The characters listed by Turco et al.[179] for separating *N. chresima* from other species in the genus could apply to several neoaplectanid species and are not at all diagnostic.

2. Bionomics

Life cycle[219] — Each female produces from 250 to 400 young. These young develop to infective stage juveniles which enter the host insect. Japanese beetle larvae exposed to soil or food containing infective stage nematodes died in 4 days.

Ecology — From 1937 to 1940, *N. chresima* was found among Japanese beetle larvae in 14 localities distributed among nine counties of New Jersey and in one locality in Maryland. [219]

3. Bacterial Associates

Glaser et al.[219] mention that dead Japanese beetle larvae parasitized by *N. chresima* assume a dirty, dull yellow color in contrast to an ocherous brown color of those parasitized by *N. glaseri.* The dull yellow color is typical of parasitism by *N. carpocapsae* and indicates the presence of a symbiotic bacterium. The authors also noted that in the presence of a nonsporulating, Gram-negative, motile bacillus, strains of *N. chresima* consistently yielded cultures on transplantation. This bacterium was probably the symbiotic form commonly associated with neoaplectanid nematodes.

4. Host Range

N. chresima was found naturally infecting corn earworms (*Helicoverpa armigera* Hub.) and Japanese beetle larvae (*Popillia japonica* Newm.). Infected experimentally were grasshoppers (*Melanoplus* sp.), larvae of the European corn borer (*Pyrausta nubilialis*), and larvae of the catalpa sphinx (*Ceratomia catalpa*).

5. Culture

After being axenized, *N. chresima* developed normally on pieces of fresh, sterile rabbit kidney placed on a nutrient agar slant. Axenic development was also achieved on an autoclaved medium composed of the following: (1) 20 g ground beef kidney or liver, (2) 100 mℓ water, (3) 0.5 g sodium chloride, and (4) 0.5 g agar. The mixture was autoclaved and could be stored for 2 weeks prior to use.

J. *Neoaplectana leucaniae* Hoy 1954

This species was described from noctuid pupae in New Zealand. Unfortunately, the description does not state whether the first or second generation adults were measured, nor is a diagnosis given.

1. Description[220]

Adults — The female tail is bluntly conical and short tipped. The head has three lips, six labial and six cephalic papillae. The pharynx has a corpus 17 μm across and metacorpus 26 μm across, an isthmus 16 μm across, and terminal bulb about 36 μm across; the latter is slightly valvated. The excretory pore is located slightly anterior to the nerve ring. The uterus contains eggs and juveniles in various stages of development. The male tail is bluntly conical, usually ending in a short tip. The head is sometimes offset. The excretory pore is located more anteriorly than in the female. Spicules are arcuate, not hooked at tip. The gubernaculum is curved posteriorly and is straight anteriorly, ending in a knob which is more strongly offset on the anterior side. One unpaired, a series of five to seven preanal subventral and five paired caudal papillae are on the tail. Quantitative values of the adults are presented in Tables 103 and 104.

Infective stage juveniles — This stage is 689 μm long and 34 μm wide. The length from the head to the nerve ring is 88 μm. The length from the head to the base of the pharynx is 109 μm. The tail is 60 μm long.

TABLE 103

Measurements of *Neoaplectana leucaniae* Females[220]

Character	Value X	Range
Total length (mm)	1.95	0.55—4.00
Greatest width	165	
Length head to base of esophagus	172	
Length head to nerve ring	117	
% vulva	56.3	
Length tail	40	
Width at anus	43	

TABLE 104

Measurements of *Neoaplectana leucaniae* Males[220]

Character	Value
Total length (mm)	0.95
Greatest width	66
Length head to base of esophagus	132
Length head to nerve ring	92
Length tail	36
Length spicule	50
Length gubernaculum	33

Diagnosis — There are very few diagnostic characters that could be used to separate this species from *N. carpocapsae* and the two may be synonomous.

2. Bionomics[220]

Life cycle — Infective stage juveniles enter the host orally with food, then develop to the adult stage in the insect's hemocoel. After mating, the female produces approximately 250 eggs. These hatch inside the female, and the young juveniles are expelled through the vulva. Eventually the remaining juveniles consume the female's body. The newly hatched juveniles molt to the second stage within 48 hr. If ample food is available, the second stage juveniles continue their development to the adult stage. If not, then they transform into "dauer" or infective stage juveniles.

Ecology — *N. leucaniae* was originally discovered in noctuid moth pupae (*Leucania acontistis* Meyr.) from a tussock base and in the larvae of a tussock moth (*Crambus simplex* Meyr.) in New Zealand. From 104 larvae of the latter insect collected, 26 contained nematodes. Hoy[220] concluded that in its natural habitat (the tussock base), *N. leucaniae* appeared to survive for long periods in the absence of host insects, yet in plot experiments, the infective juveniles disappeared after 3 months. The lack of persistence in the soil limited the effectiveness of *N. leucaniae* as a biological control agent against soil insects. However, out of 70 adult beetles of *Costelytra zealandica* that died when confined with infective stage juveniles, 22 were infected with *N. leucaniae*. Of 30 beetles that flew out of the experimental plot, 3 contained nematodes. This shows how the nematode can be spread by adult insects.

3. Bacterial Associates

There was no indication or evidence of a bacterium associated with *N. leucaniae*, although there is no reason to believe why one would not be present.

TABLE 105

Host Range of *Neoaplectana leucaniae*

Order	Family	Host	Stage	Infection[a]
Coleoptera	Scarabaeidae	*Costelytra zealandica* (White)	L	E
		Odontria communis Given	L	E
		O. nitidula Brown	L	E
		O. autumnalis Given	L	E
		O. striata White	L	E
		Pyronota festiva (Fab.)	L	E
		P. inconstans Brookes	L	E
Lepidoptera	Noctuidae	*Leucania acontistis* Meyr.	P	N
	Pyralidae	*Crambus simplex* Meyr.	L	N

[a]   E, experimental infection; N, natural infection.

After Hoy, J. M., *Parasitology*, 44, 392, 1954.

TABLE 106

Measurements of *Neoaplectana janickii* Females[221]

Character	Value
Total length (mm)	0.940—1.680
Greatest width	94—152
Length stoma	8
Length head to base of esophagus	177
Length head to nerve ring	115
% vulva	58.6
Length tail	92
Width at anus	39

TABLE 107

Measurements of *Neoaplectana janickii* Males[221]

Character	Value
Total length (mm)	0.996—1.283
Greatest width	48—80
Length stoma	9
Length head to base of esophagus	151
Length head to nerve ring	113
Length tail	40
Width at anus	43
Length spicule	42
Width spicule	16
Length gubernaculum	24
Length tail projection	4—6

4. Host Range

The host range of *N. leucaniae* as reported by Hoy[220] is given in Table 105.

5. Culture

N. leucaniae was tested against insects in laboratory experiments only. Hoy[220] was unsuccessful in attempting to culture the nematode on veal infusion agar or nutrient agar slants with sterile rabbit, sheep, or beef kidney. However, *N. leucaniae* was cultured on the veal pulp medium used for *N. glaseri* by McCoy and Girth.[210] The capacity of the nematode to parasitize insects was not hindered after being cultured on artificial media for ten generations.

K. *Neoaplectana janickii* Weiser and Kohler 1955

This neoaplectanid was described from the sawfly *Acantholyda nemoralis* Thoms. in Silesia, Poland. Aside from the original description in 1955,[221] no further mention of this species has been given.

1. Description[221]

Adults — Females vary in length. The tail has a small spike, and there are three distinct lips, each with a bristle. The vulva is slightly protruding. Quantitative values are given in Table 106. The male tail terminates in a small spike. There are four pairs of copulatory papillae. Quantitative values are given in Table 107.

Diagnosis — Since the presence of lip bristles and the number of copulatory papillae presented in the original description are questionable characters and should not be used in a diagnosis, it is difficult to separate this species from other members of the genus.

2. Bionomics and Ecology
The original outbreak of *N. janickii* occurred in fir forests of Dabrawa Opolska in Silesia.[221] This represents a new ecological habitat for a neoaplectanid nematode. The infection of *Acantholyda nemoralis* reached 42% in Dabrawa Opolska.

3. Bacterial Associates
Weiser and Köhler[221] mentioned that insect larvae infected with *N. janickii* were slightly darkened in color, but not decayed, and of roughly the same consistency as healthy larvae. These observations suggest the presence of a symbiotic bacterium.

4. Host Range
Only the above-mentioned pamphiliid sawfly was cited as a host for this nematode.

L. *Neoaplectana carpocapsae* Weiser 1955
This species was described by Weiser[222] from codling moth larvae collected in Czechoslovakia in 1954. At the same time, Dutky and Hough[223] reported finding a similar species in codling moth larvae collected in the Eastern part of the U.S. There has been some confusion regarding the taxonomic position of these two nematodes, especially since the American material was known for a period only by the code DD-136. Since hybridization experiments showed that the Czechoslovakian and American forms would cross,[188] they were treated as strains of *N. carpocapsae*. The population originally described by Weiser was labelled the Czechoslovakian strain and the American population was labelled the DD-136 strain. Since then, other strains of *N. carpocapsae* have been recovered from different geographic localities.

1. Description
Czechoslovakian strain — Aside from the original description of this nematode by Weiser,[222] Poinar,[188] Turco et al.,[179] and Stanuszek[183] have provided subsequent descriptions which will also be cited here. They show how much variability can occur in the same species of *Neoaplectana* when examined under different conditions.

Adults — According to Weiser,[222] the female head ends with three lips bearing papillae. The pharynx has an anterior cylindrical portion, followed by an isthmus and a terminal bulb with valvular apparatus. The rectum is located about 40 to 50 m*l* below the end of the gonads. The vulva protrudes slightly. In males, the tail has a variable number of genital papillae. Spicules are equal in size, tips pointed, with a ventral arch. The gubernaculum is long and flat with the distal portion narrowed. Quantitative values of the adults are presented in Tables 108 and 109

Juveniles — According to Weiser,[222] two types are noted, those up to 500 μm (first stage) and those over 500 μm (second stage). The following measurements of the second (actually, the third) stage juveniles are given: total length, 720 μm; length of pharynx, 84 μm; greatest width, 28 μm; length of tail, 88 μm.

Weiser did not indicate any character which distinguished his *N. carpocapsae* from previously described species in the genus.

Adults — According to Poinar,[188] the cuticle is smooth, the head truncate to slightly rounded, and the lips united. Seate are obscure. There are two circles of anterior papillae, six inner labial papillae, and six outer cephalic papillae. Amphids are small and porelike and are near the level of the cephalic papillae. The stoma is partially collapsed

TABLE 108

Measurements of *Neoaplectana carpocapsae* Females[222]

Character	Value — first generation	
	X̄	Range
Total length (mm)	1073	(084—1610)
Greatest width	94	
Length stoma	6	
Width stoma	17	
Length head to base of esophagus	175	
Length head to nerve ring	100	
% vulva	55	(38—65)
Length tail	53	
Width at anus	43	

TABLE 109

Measurements of *Neoaplectana carpocapsae* Males[222]

Character	Value — first generation	
	X̄	Range
Total length (mm)	643	(529—708)
Greatest width	43	
Length stoma	4	
Width stoma	11	
Length head to base of esophagus	120	
Length head to nerve ring	61	
Length tail	24	
Width at anus	34	
Length spicule	42—60	
Width spicule	6—8	
Length gubernaculum	40—42	

with only an anterior vestibule remaining. A collar is lacking. Pharyngeal tissue is close to the mouth opening, reaching to the base of the vestibule. Cheilorhabdions are represented as lightly sclerotized areas lining the inside of the lip region anterior to the pharynx. A small sclerotized area just beneath the cheliorhabdions probably represents the modified prorhabdions. Meso-, meta-, and telorhadions are vestigial, although sometimes rarely represented as refractive edges lining the collapsed walls of the stoma. The pharynx is muscular, and the anterior portion of the procorpus slightly expanded just behind the vestibule, then extends into a slightly enlarged nonvalvated metacorpus, followed by an isthmus and terminates in a basal bulb containing a small haustrum with three bulb flaps lined with refractive ridges. The base of the pharynx is often inserted into the anterior portion of the intestine. The nerve ring surrounding the isthmus is just anterior to the basal bulb. The excretory pore is usually anterior to the nerve ring. The lateral field and phasmids are inconspicuous. The female is amphidelphic with opposed reflexed ovaries. It is variable in size, some giant forms reaching 10 mm. The vulva is a transverse slit, bearing two prominant ventral protuberances. The vagina is short with muscular walls leading into a prouterus which serves as an egg chamber — a small constriction separates this from the remainder of the uterus, where fertilization occurs. A well-developed glandular oviduct leads into the growth zone of the ovary and finally into the elongate germinal zone. The female tail is bluntly conical to dome-shaped, with or without a short spine on the tip. Second and succeeding generation females in the host are correspondingly smaller in size. Pigmy forms or swollen miniature females were never found. The male has a single reflexed testis consisting of a germinal and growth zone leading into a seminal vesicle containing spermatophores. The vas deferens is conspicuous, with glandular walls. Spicules are paired, symmetrical, curved, and bear a more or less pronounced arch on their ventral surface. The shape of the capitulum is variable, from slightly pointed to round or flat. The surface and edge of the calamus and lamina bear ridges. A thin velum is present. The gubernaculum is also variable, ranging from completely flattened to bowshaped in lateral view, with the proximal portion bent at various angles and sometimes even bluntly bifurcate. In dorsal view, the distal portion consists of two lateral projections with a thin sclerotized spine between them. The area between the lateral projections is connected by a thin membrane. The male tail has a complement of 23 anal papillae (11 pairs and a single median adanal) comprising two rows of six ventrolateral papillae and five paired post anal papillae. Of the latter group, two pair are situated laterally on the tail, one near the terminus and the other in the vicinity of the gubernaculum. This latter pair is variable in position and often difficult to observe. The tip of the tail is conical with a small appendage. A bursa is absent.

Infective stage (third stage) juveniles —(for illustration, see Poinar)[224] They are much narrower than the corresponding parasitic juvenile. The mouth and anal opening are closed. The pharynx and intestine are collapsed. The tail is pointed, lateral fields are distinct (Figures 41 and 42). The infective third stage juveniles possess conspicuous lateral fields. The number and shape of the folds and striae constituting these fields are sometimes characteristic and used in the differentiation of nematode species. Electron micrographs show that these fields at midbody consist of six longitudinal ribs, comprising two pairs of pronounced outer ribs on either side of a pair of finer inner ribs.

Smears of testicular tissue showed a chromosomal condition of four bivalents and a single univalent. This indicates a 2N condition of nine chromosomes for the males and ten for the females. Quantitative measurements of the Czechoslovakian strain are presented in Tables 110 through 115.

The nematode known today as *N. carpocapsae* is widely distributed in nature and has been collected in different geographical areas. For the practical purpose of keeping the various geographical populations separate for experimental reasons, the term "strain" will be used here to designate the various populations of *N. carpocapsae*. Each strain refers to the descendents of a single isolation collected from the field. Some of the common strains are listed in Table 116. There have been no qualitative differences found yet between the various strains of *N. carpocapsae*, and more work is needed to evaluate quantitative measurements for characteristic differences. However, because of their importance in the area of biological control, quantitative measurements of the DD-136 and *Agriotos* strains of *N. carpocapsae* are presented in Tables 117 to 123. It should be stressed that quantitative differences may even occur between different isolates of the same strain, as shown in the measurements of the Czechoslovakian and DD-136 strains made separately by Stanuszek (Tables 113 to 115 and Tables 120 and 121) and Poinar (Tables 110 to 112 and 117 to 119), respectively.

Although these strains are not genetically isolated, as determined by interbreeding studies in the laboratory, they may differ from each other in physiological qualities. Thus, Hansen et al.[225] reported that in axenic liquid media distributed as a film over glass wool, very few infective stages of the Czechoslovakian strain exsheathed and reached the adult stage, whereas most of the DD-136 strain reached the adult stage and reproduced. Thus, one of the strain differences appears to be the ability of the infective stages to exsheath and initiate development when placed in liquid media.

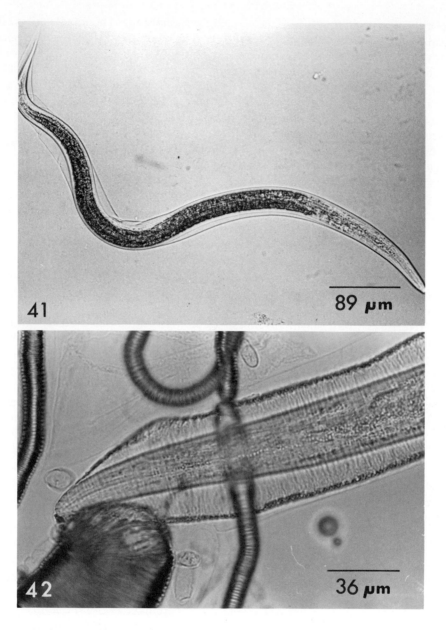

FIGURE 41 and 42.   (41) Third-stage infective juvenile of *Neoaplectana carpocapsae* enclosed in its second stage cuticle. (42) Infective juvenile of *Neoaplectana carpocapsae* in the trachea of a California oak moth.

2. Bionomics

Life cycle — The infective stages of *N. carpocapsae* normally enter their hosts by way of the mouth. Studies of parasitism of the greater wax moth larvae (*Galleria mellonella*) by *N. carpocapsae* showed that the nematodes occurred first in the crop and midgut of the host.[229] It was concluded that the nematodes entered the alimentary tract via the mouth. The ensheathing second stage cuticle was shed in the crop or midgut, and the nematode began burrowing between the midgut epithelial cells toward the gut serosa and hemocoel. The nematodes then penetrated the gut wall (presumably by mechanical pressure) and were found in the hemocoel about 11 hr after entry. Soon afterwards, these nematodes underwent changes leading to the formation of the para-

TABLE 110

Comparative Measurements of Females of the Czechoslovakian Strain of *Neoaplectana carpocapsae* Weiser[188]

Character (N = 21)	Value	
	$\overline{X}$	Range
Total length (mm)	3.31	1.97—5.81
Greatest width	156.3	123.2—192.5
Stoma length	5.9	3.1—7.8
Stoma width	9.3	6.2—12.4
Length head to excretory pore	56.7	34.1—86.8
Width at excretory pore	82.2	65.1—108.5
Length head to nerve ring	140.1	108.5—170.5
Length head to base of esophagus	116.9	91.0—133.3
Width at base of esophagus	201.5	155.0—241.8
Length tail	131.4	111.6—145.7
Width tail	34.7	18.6—55.8
% vulva	52.0	50.0—58.0

TABLE 111

Comparative Measurements of Males of the Czechoslovakian Strain of *Neoaplectana carpocapsae* Weiser[188]

Character (N = 12)	Value	
	$\overline{X}$	Range
Total length (mm)	1.32	1.25—1.40
Greatest width	117.0	107.8—130.9
Stoma length	5.6	5.2—7.8
Stoma width	4.6	2.6—6.5
Length head to excretory pore	71.3	55.8—74.4
Width at excretory pore	52.1	46.5—58.9
Width at nerve ring	127.7	108.5—155.0
Length head to base of esophagus	151.9	145.7—155.0
Width at base of esophagus	70.7	65.1—77.5
Length bent portion of testis to base of esophagus	136.3	84.7—192.5
Length tip of testis to tail	1000.0	780.0—1090.0
Reflection testis	385.0	284.9—477.4
Length tail	31.2	22.1—35.1
Width anus	44.5	39.0—52.0
Length spicula	66.3	62.4—70.2
Width spicula	12.5	10.4—13.0
Length gubernaculum	47.1	42.9—52.0
Width gubernaculum	6.4	5.2—7.8

sitic juvenile stages. The body enlarged, the pharyngeal bulb started pulsating, and the alimentary tract opened. The developing third and fourth stage juveniles invaded the fat body and malpighean tubules, and about 18 hr after entry into the gut they could be found in the silk glands and muscles. At this time the gut epithelium and fat body began to disintegrate.

The host generally dies about 24 hr after the nematodes enter the alimentary tract, yet the nematodes continue to develop into first generation adults which are large and robust (Figure 43). These adults mate, and the female initially deposits eggs, but later

TABLE 112

Measurements of the Infective Stages of the Czechoslovakian Strain of *Neoaplectana carpocapsae*[188]

Character	Value X	Value Range
Total length (mm)	572.0	488.0—613.0[2]
Greatest width	26.0	25.0—28.0
Head to excretory pore	42.1	39.0—56.0[3]
Head to excretory pore (sample 2)	43.2	39.0—58.5[3]
Head to nerve ring	88.0	84.0—93.0[2]
Length tail	53.0	47.0—59.0

TABLE 113

Measurements of the Males of the Czechoslovakian Strain of *Neoaplectana carpocapsae*[183]

Character	First generation X	First generation Range	Second generation X	Second generation Range
Total length (mm)	1.73	1.50—2.75	1.13	0.72—1.58
Greatest width	136	80—235	58	43—61
Length stoma	6	4—7	5	3—8
Width stoma	4	3—7	6	3—8
Length head to base of esophagus	190	170—200	142	131—156
Length head to excretory pore	82	61—108	39	32—43
Length head to nerve ring	178	134—190	105	85—116
Length reflexion of testis	721	414—967	589	350—946
Length tail	47	36—50	35	27—42
Width at anus	51	46—55	44	34—47
Length spicule	85	69—93	66	55—73
Length gubernaculum	52	46—57	45	39—50
Length mucron	3	2—4	3	2—4

TABLE 114

Measurements of the Females of the Czechoslovakian Strain of *Neoaplectana carpocapsae*[183]

Character	First generation X	First generation Range	Second generation X	Second generation Range
Total length (mm)	10.99	7.12—16.98	2.405	1.304—3.550
Greatest width	325	215—391	172	95—189
Length stoma	10	6—11	6	4—10
Width stoma	15	10—15	7	5—12
Length head to base of esophagus	310	225—382	221	156—279
Length head to excretory pore	119	87—157	80	56—98
Length head to nerve ring	248	211—279	178	129—221
% vulva	54	49—56	53	50—57
Length tail	76	52—100	35	28—50
Width at anus	168	110—253	63	39—83
Length tail knob	10	9—16	10	9—21

TABLE 115

Measurements of Infective Stage Juveniles of the Czechoslovakian Strain of *Neoaplectana carpocapsae*[183]

	Value	
Character	$\overline{X}$	Range
Total length (mm)	572	498—650
Greatest width	23	20—30
Length head to excretory pore	42	38—51
Length head to nerve ring	87	80—99
Length head to base of pharynx	115	103—190
Length tail	50	46—61
Width tail	14	10—16

TABLE 116

Strains of *Neoaplectana carpocapsae* Weiser

Strain	Original host	Host order	Geographical area	Ref.
Czechoslovakian	*Laspeyresia pomonella* (L.)	Lepidopter	Czechoslovakia	222
DD-136	*L. pomonella* (L.)	Lepidoptera	Virginia, U.S.	223
Mexican	*L. pomonella* (L.)	Lepidoptera	Alende, Chihuahua, Mexico	226
Agriotos	*Agriotes lineatus* L.	Coleoptera	Leningrad, U.S.S.R.	189
All	*Vitacea polistiformis* (Harris)	Lepidoptera	Georgia, U.S.	227
X-I	*Laspeyresia pomonella* (L.)	Lepidoptera	Poland	183
X-III	*Agrotis segetum* (Schiff)	Lepidoptera	Poland	183
X-IV	*Pieris brassicae* (L.) *Barathra brassicae* L.	Lepidoptera	Poland	183

the eggs hatch in her body. Some of these eggs (depending on the initial dosage) will produce juveniles which develop all the way to second generation adults. Because of lack of food and crowding, these adults (especially the females) are generally much smaller than the first generation adults. After mating, the second generation adults deposit eggs which produce juveniles which generally form infective stages. The infective stages are again third stage nematodes which are enclosed in second stage cuticles. They are morphologically and physiologically different from developing third stage parasitic juveniles. The mouths of the infective stage juveniles are closed and the alimentary tract is collapsed and nonfunctional.[230] These nonfeeding juveniles can survive for long periods in the absence of hosts and are the only stage which can initiate infection of an insect. These stages leave the host cadaver in large numbers and enter the environment where they then make contact with another host.

Infection of mosquito larvae by way of the mouth was discussed by Welch and Bronskill.[231] Weiser[188] reported that *N. carpocapsae* could enter the body cavity of some insects via the spiracles and trachea, and Triggiani and Poinar[232] showed that infective stages of *N. carpocapsae* could gain access to the body cavity of adult Lepidoptera through the spiracles and trachea (Figure 42). Poinar et al.[233] showed that the infective stages could also enter the pneumastic lobes of tsetse fly puparia and undergo development (Figure 49) Schmiege[234] also noted that pupae of the cerambycid beetle *Monochamus scutellatus* Say were successfully attacked by *N. carpocapsae*.

There are four juveniles stages during the development of *N. carpocapsae* which is the normal number for nematodes. The "dauer" or infective juvenile is not an oblig-

TABLE 117

Measurements of Females of the DD-136 Strain of *Neoaplectana carpocapsae* Weiser[188]

Character	Value — first generation	
	$\overline{X}$	Range
Total length (mm)	3.68	2.80—5.16
Greatest width	148	123—185
Length stoma	7	5—9
Width stoma	8	6—9
Length head to base of esophagus	191	161—217
Length head to excretory pore	61	47—74
Length head to nerve ring	137	118—161
% Vulva	54	52—56
Length tail	36	28—46
Width at anus	69	50—87

TABLE 118

Measurements of Males of the DD-136 Strain of *Neoaplectana carpocapsae* Weiser[188]

Character	Value — first generation	
	$\overline{X}$	Range
Total length (mm)	1.45	1.09—1.71
Greatest width	101.6	77.0—130.9
Stoma length	5.6	2.6—7.8
Stoma width	4.9	3.9—6.5
Length head to excretory pore	61.4	46.5—74.4
Width at excretory pore	47.7	37.2—58.9
Length head to nerve ring	110.1	93.0—124.0
Width at nerve ring	58.9	46.5—71.3
Length head to base of esophagus	154.7	136.4—167.4
Width at base of esophagus	64.5	49.6—77.5
Length bent portion of testis to base of esophagus	128.6	53.9—284.9
Length tip of testis to tail	1150.0	780.0—1560.0
Reflection testis	563.6	400.4—808.5
Length tail	30.4	23.4—39.0
Width anus	42.6	32.5—54.6
Length spicula	64.6	58.5—71.5
Width spicule	11.1	9.1—13.0
Length gubernaculum	47.1	39.0—55.9
Width gubernaculum	5.2	3.9—6.5

atory stage but one that is produced as a result of crowding, lack of food, or unfavorable environment. This stage is also resistant to external conditions and can survive for long periods in the soil. They appear to actively "search" for insect hosts by standing on their tails and waving back and forth. Their unique ability to "jump" and increase their searching ability has been described by Reed and Wallace.[235]

Ecology — Because of the biological control potential for *N. carpocapsae,* investigations were conducted to determine the effect of the physical environment on the infective stages.

Temperature — Schmiege[234] placed infective stage nematodes in water drops held at various temperatures and found that an exposure of 1 hr at 35°C caused high mortal-

TABLE 119

Measurements of the Third Stage Infective Juveniles of the DD-136 Strain of *Neoaplectana carpocapsae* Weiser[188]

Character	Value	
	X̄	Range
Total length (mm)	547.0	438.0—625.0
Greatest width	24.0	22.0—28.0
Head to excretory pore	35.7	34.0—40.0
Head to excretory pore (sample 2)	38.6	36.4—40.3
Head to nerve ring	85.0	81.0—90.0
Length tail	53.0	50.0—59.0

TABLE 120

Measurements of Males of the DD-136 Strain of *Neoaplectana carpocapsae* Weiser[183]

	Value			
	First generation		Second generation	
Character	X̄	Range	X̄	Range
Total length (mm)	1.89	1.65—2.89	1.25	0.58—1.50
Greatest width	129	70—230	65	50—71
Length stoma	61	4—7	4	2—7
Width stoma	4	4—6	5	3—8
Length head to base of esophagus	191	167—205	150	134—160
Length head to excretory pore	82	62—109	47	35—72
Length head to nerve ring	169	133—197	98	89—117
Length reflexion of testis	682	426—1052	605	356—920
Length tail	41	34—51	36	25—39
Width at anus	47	40—55	45	34—60
Length spicule	87	71—100	63	56—70
Width spicule	13	11—15	10	9—12
Length gubernaculum	51	46—69	46	40—50
Length mucron	3	2—4	3	2—4

TABLE 121

Measurements of Females of the DD-136 Strain of *Neoaplectana carpocapsae* Weiser[183]

	Value			
	First generation		Second generation	
Character	X̄	Range	X̄	Range
Total length (mm)	10.67	9.56—17.40	2.55	0.70—4.60
Greatest width	351	225—410	146	58—216
Length stoma	10	5—10	6	3—9
Width stoma	12	10—14	7	6—10
Length head to base of esophagus	306	226—389	200	157—262
Length head to excretory pore	122	88—157	70	55—96
Length head to nerve ring	245	213—275	159	125—183
% vulva	56	50—57	55	52—58
Length tail	68	53—81	39	28—47
Width at anus	115	97—115	58	41—77
Length tail knob	10	9—20	5	3—7

TABLE 122

Measurements of the Females of the *Agriotos* Strain of *Neoaplectana carpocapsae* Weiser[228]

Character (N = 10)	Value	
Total length (mm)	6.30	(5.76—7.14)
Greatest width	1205	
Length stoma	21	(16—24)
Length head to base of esophagus	192	(176—203)
Length head to excretory pore	70	
Length head to nerve ring	170	
% vulva	48.7	(48—49.9)
Length tail	32	(27—43)
Width at anus	68	
Length tail knob	2.4	

TABLE 123

Measurements of the Males of the *Agriotos* Strain of *Neoaplectana carpocapsae* Weiser[228]

Character (N = 10)	Value	
Total length (mm)	1.45	(1.33—1.57)
Greatest width	120	
Length stoma	14	
Length head to base of esophagus	150	(146—157)
Length head to excretory pore	68	
Length reflexion of testis	100	
Length tail	43	
Width at anus	100	
Length spicule	57	(56—65)
Width spicule	11	(11—13)
Length gubernaculum	44	(40—48)
Length tail spike	3.6	

ity, and those that survived the treatment never resumed normal activity. Exposure for 16 hr at 37°C and 1 hr at 41°C resulted in 100% mortality. Thus, Schmiege pointed out that this nematode could not become established in most warm-blooded animals, including man. However, the infective stage juveniles can be he held in water at 50°C for up to several years with only a gradual loss of infectivity. Schmiege[234] also noted that whereas 70% of the infective stages (N = 28,000 nematodes) were killed when placed in a freezer and kept at −10°C for 18 hr the survivors were still infective for insects. Tests also showed that the nematodes could overwinter in the soil at the latitude of St. Paul, Minnesota. As far as movement was concerned, the temperature range from 15 to 28°C seemed to be most favorable.

The effect of temperature on the growth and reproduction of DD-136 was studied by Kaya.[236] This author found that the most favorable temperature for development was between 23 and 28°C. No growth was observed below 10°C or above 33°C. On the basis of these studies, Kaya concluded that the likelihood of this nematode developing in birds and mammals was remote.

Moisture — The moisture requirement for most nematodes is well known, and dessication is one of the most important factors limiting the action of neoaplectanid nematodes. Dutky[237] mentioned that *N. carpocapsae* was not resistant to drying and was quickly killed by desiccation. Schmiege[234] showed that under conditions approaching

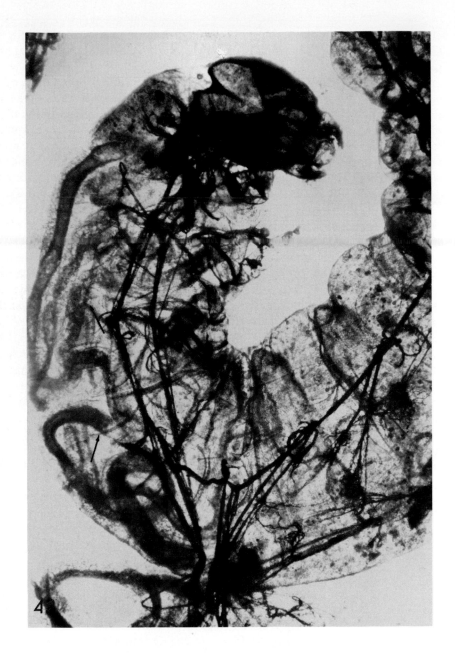

FIGURE 43.   First generation adults of *Neoaplectana carpocapsae* (arrows) inside a corn root-worm larva (*Diabrotica* sp.).

100% relative humidity, the infective stages of *N. carpocapsae* survived indefinitely on depression slides. At a relative humidity of 96 to 97% (at 23°C), the infective juveniles survived only for a little over 3 hr. Welch and Briand[238] found that 50% of the infective stages of *N. carpocapsae* died after 26 min at 80% relative humidity (RH) and 38 min at 90% RH. Kamionek et al.[239] placed infective stages of *N. carpocapsae* on *Tradescantia* leaves which were placed immediately afterwards in an environmentally controlled chamber. They showed that at 30°C 98% mortality occurred in 2.5 hr at 20% RH and after 102 hr at 85% RH. At 5°C, 98% mortality occurred after 24 hr at 20% RH and after 36 hr at 85% RH. However, it is now obvious that under a more natural habitat, such as in soil, the infective stages of *N. carpocapsae* can withstand relatively low relative humidities. Moore[240] showed that when a quart of moist soil containing *N. carpocapsae* infectives dried slowly at 70% RH, the nematodes were active until the 20th day.

Simons and Poinar[241] demonstrated how important the gradual desiccation of the nematodes was for survival. If infective stages are placed in a drop of water on a cover slide and then examined several minutes after the water has evaporated, all the nematodes will be dead. If, however, that drop is placed in chambers of decreasing amounts of RH over 12-hr periods, then the dried slide with nematodes can be placed for several days at room temperature, and over 50% of the nematodes will survive. This prolonged dying is exactly what happens in nature and reflects the true ability of *N. carpocapsae* to resist drought conditions in the soil. Using this technique, Simons and Poinar showed that 90% of the nematodes were still viable at 79.5% RH after 12 days. This RH corresponds to a pF of 5.5 and is well below the permanent wilting point of plants. Even after 4 days, exposure to 48.4% RH (pF 6.0), which corresponds to very dry soil, more than 80% of the nematodes survived. These results show that the infective stages of *N. carpocapsae* can survive desiccation for relatively long periods, even when the RH falls well below the wilting point of plants. Visually, the desiccated nematodes are immobile, collapsed, twisted, and for all practical purposes appear dead; yet they readily revive when immersed in water overnight.

Various authors have attempted to retard evaporation when applying aqueous suspensions of *N. carpocapsae* infective stages to above-ground surfaces. Welch and Briand[238] tested glycerin, honey, glucose, sorbitol, urea, and agar, but found that they were effective only at high concentrations. However, at such concentrations, the chemicals were often phytotoxic or nematicidal or they enhanced fungal disease. Webster and Bronskil[242] achieved some success in prolonging the life of infective stage *N. carpocapsae* on foliage by adding a mixture of the water thickener Gelgard M (0.13%), the evaporation retardant Folicote 351 (0.2%), and a surfactant (Arlatone T, 0.1%) to the nematodes. Under these circumstances, the average mortality of larch sawfly larvae [*Pristiphora erichsonii* (hartig)] was increased from 24 to 90%.

More recently, Bedding[243] discovered that the infective stages of *Neoaplectana* spp. could live for long periods without free water when placed in a variety of inert organic solvents including paraffin oils. Spraying nematodes on plants in a paraffin-oil mixture allowed them to remain viable for up to 2 months and become active when ingested by insects.

Migration in soil — When infective stages of *N. carpocapsae* are mixed in a small amount of soil, most soon migrate to the surface of the soil and stand on their tails (Figure 44). This characteristic movement toward the surface seems to be a natural feature of most strains of *N. carpocapsae*. Reed and Carne[244] noted this behavior and stated that the DD-136 strain failed to penetrate deeply into soil. The above authors described three types of movement in the infective stages: gliding, bridging, and leaping. Gliding movement is described as the typical worm-like movement characteristic of nematodes in general and used by DD-136 to reach the soil surface. After reaching the soil surface, bridging movement ensued. This consisted of the nematodes standing erect and waving their anterior ends. If the head made contact with another soil particle, the nematode drew itself up on the particle. This behavior took large numbers of nematodes to elevated positions. Leaping movement consisted of the nematode coiling around a water droplet, then suddenly uncoiling and being projected horizontally up to 10 mm across the substrate. Reed and Carne[244] concluded that the DD-136 strain is adapted for movement on the soil surface and prefers the surface to a subterranean environment.

El-Sherif[244a] tested the vertical migration of the DD-136 and *Agriotos* strains of *N.*

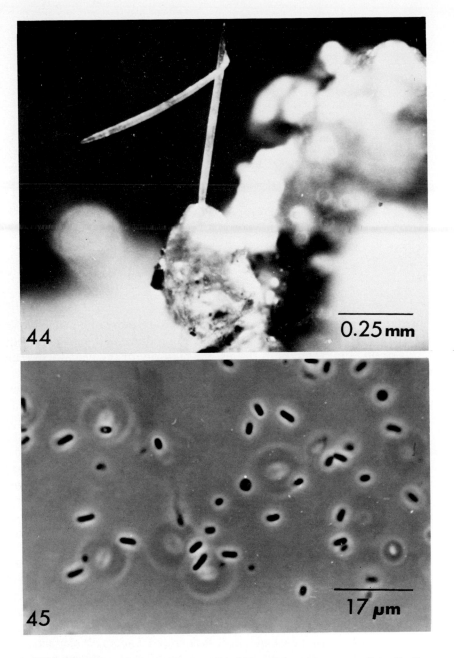

FIGURE 44 and 45.　　(44) Infective juvenile of *Neoaplectana carpocapsae* "standing" on a soil particle with another juvenile attached. (45) Cells of *Achromobacter nematophilus* in the blood of an insect recently attacked by *Neoaplectana carpocapsae*.

carpocapsae in columns of sandy loam soil. Each column was 30 cm high and 7.2 cm in diameter and was composed, of six sections. (Each section was 5 cm high.) The nematode inoculum (infective stage juveniles) was placed in the middle of the column. After the nematodes were kept in the soil column for a specified time, each section of the column was removed, and the nematodes were extracted with sieves and a modification of the Baermann funnel technique. The total number of nematodes recovered from each section of soil column is recorded in Table 124.

TABLE 124

Vertical Migration of the Infective Stages of the
DD-136 and *Agriotos* Strains of *Neoaplectana*
carpocapsae in a 30-cm-high Soil Column[244a]

Soil section (cm)	Nematodes recovered (living)	
	DD-136	*Agriotos* strain
25—30	980	40
20—25	590	10
15—20	1583	140
10—15	996	219
5—10	580	63
0—5	70	10

Note: Nematodes were added to the soil in the
middle of the column and extracted 15
days later (age of nematodes — 30 days).

These preliminary results show that (1) there tends to be more of a migration toward the soil surface in the DD-136 strain, (2) the migration down and up is about equal in the *Agriotos* strain, and, (3) in both strains, the majority of the nematodes remained in the center of the column.

Sunlight and ultraviolet radiation — It was recently shown that the infective stage juveniles of *N. carpocapsae* were very sensitive to both short UV light (254 nm) and natural sunlight.[244b] Although irradiation with UV light did not appear to alter movements of the juveniles, inhibition of development and reproduction occurred at exposure periods under 7 min. At longer exposures, the nematodes were unable to cause lethal infections and eventually died. A 95% loss of pathogenicity also occurred when the juveniles were exposed to 60 min of direct sunlight. These findings demonstrate how important UV radiation is and that it should be taken into consideration when using these nematodes in the field.

3. Bacterial Associates

The presence of bacteria associated with *N. carpocapsae* was first mentioned by Weiser[222] for the Czechoslovakian strain, then discussed by Dutky[237] for the DD-136 strain. Poinar and Thomas[245] noted a specific bacterium which was consistently present inside the majority of infective juveniles of the DD-136 strain and was found inside the body cavity of recently infected insect hosts (Figure 45). Since neither Weiser or Dutky described the bacteria they had noted, Poinar and Thomas[245] characterized and described their bacterium as *Achromobacter nematophilus*. A supplementary description of this bacterium and a similar one isolated from the *Agriotos* strain of *N. carpocapsae* was subsequently published.[246] Characteristics of *A. nematophilus* (based on the above studies) are presented in the section on associated microorganisms.

A. nematophilus is carried in the lumen of the ventricular portion of the infective stage juveniles of *N. carpocapsae*.[224] When these nematodes are placed in a drop of blood or penetrate into the body cavity of an insect host, they initiate normal development, and the bacteria are released through the anus and multiply rapidly in the host's body. In two instances, bacterial cells resembling *A. nematophilus* in form were found inside the anterior intestinal cells of infective stage nematodes.

A. nematophilus has never been recovered in nature except in the body of insects attacked by certain neoaplectanid nematodes and inside the intestine of the developing and infective stage nematodes. The bacteria are protected inside the nematode between periods of active infection since the cells are very fastidious and have a very transient existence in soil and water.

Poinar and Thomas[247] showed that from one to three cells of *A. nematophilus* are pathogenic when experimentally introduced into the hemolymph of *Galleria* larvae. However, large numbers of the bacteria have no effect when introduced into the insect's alimentary tract.

When cells of *A. nematophilus* are introduced into the body cavity of *Galleria* larvae, the insects stop feeding after several hours and death usually occurs within 48 hr. At this time, the cadaver becomes a characteristic creamy-yellow in color, which is typical of neoaplectanid infections. Insect death probably arises from the production of proteolytic enzymes which may explain the relative lack of resistance to *A. nematophilus* when just one to three cells are injected into the body cavity of *Galleria* larvae.

The significance of the association between *N. carpocapsae* and *A. nematophilus* was studied by Poinar and Thomas.[205] By culturing the host insect (*Galleria mellonella*), *N. carpocapsae,* and *A. nematophilus* separately under axenic conditions, it was possible to study the association between the nematode and bacterium. It was shown that axenic infective stages were able to penetrate and enter the body cavity of *Galleria* without the bacterium, but were unable to reproduce. It was shown earlier that *A. nematophilus* is very fastidious and cannot infect the insect when placed in the digestive tract. Thus, the nematode protects the bacterium in nature and carries it into the hemocoel of a new host, while the bacterium furnishes some nutrient, either directly or indirectly, that is necessary for nematode reproduction. Clearly, the association is a case of mutualism.

The present author conducted several tests to determine under what nutritional conditions *N. carpocapsae* could reproduce. When axenic nematodes were placed on nutrient agar plates seeded with *A. nematophilus,* the nematodes developed to the adult stage, mated, and produced eggs. These eggs developed to a smaller second generation of adults that died without mating or producing any eggs. Transferring these second generation adults to new nutrient agar (NA) plates did not result in further development. Tests were conducted by placing axenic infective juveniles in various mixtures of *Galleria* hemolymph, basal defined medium, soy peptone and yeast extract, heated liver extract, defined medium with protein added, cells of *A. nematophilus,* folic acid, vitamins, glutamic acid, amino acids, yeast extract, and lactalbumin hydrosate (Table 125). These tests show that there is no simple substance that can be added to insect hemolymph to replace *A. nematophilus* for normal nematode development.

The question now arises as to whether the nutritive factors are living bacterial cells or a breakdown product as a result of bacterial activity. *Galleria* larvae that had been injected with *A. nematophilus* were ground up and the mixture passed through a filter to remove all bacteria. When the filtrate was used to culture the infective stages, there was development and limited reproduction. This shows that products produced by the bacteria contain enough nutrients for some reproduction, although this material is probably more concentrated in living bacterial cells.

There are additional points of interest regarding insect mortality from *Neoaplectana* infection. It is true that under natural conditions insect mortality from *Neoaplectana* infection is due to the action of *A. nematophilus.* However, Poinar and Thomas[205] showed that one axenic nematode in an insect could eventually cause the host's death. It was speculated that in such a case, insect mortality was probably associated with the release of material from the developing nematode which could act as a toxin. Indeed, when liquid media which supported the growth of two strains of *N. carpocapsae* and *N. glaseri* were injected into the body cavity of *Galleria,* many insects died (Table 126). Unused media produced no reaction. The developing axenic nematodes produced some product which was clearly toxic to *Galleria* larvae.

TABLE 125

Growth of Axenic Infective Stage Juveniles In Various Combinations of Nutrients

	Development		
Combination	Postdauer	Adults	Reproduction
Heated liver extract (HLE)	+	−	−
HLE plus *Galleria* hemolymph	+	+	−
Defined medium with protein	−	−	−
HLE plus *G.* hemolymph	+	+	−
HLE plus basal defined medium (BDM)	+	+	−
Soy peptone and yeast plus hemolymph	+	+	Eggs, but no young
BDM plus hemolymph	+	+	−
Soy peptone-yeast plus *A. nematophilus*	+	+	Limited
Soy peptone plus HLE plus *A. nematophilus*	+	+	+
Hemolymph plus folic acid	+	+	−
Hemolymph plus *A. nematophilus*	+	+	+
Hemolymph plus vitamins	+	+	−
Hemolymph plus folic acid	+	+	−
Centrifuged hemolymph with *A. nematophilus*	+	+	−
Hemolymph	+	+	−
Hemolymph and vitamins	+	+	−
Hemolymph and amino acids	+	+	−
Hemolymph and glutamic acid	+	+	−
Hemolymph with vitamin and amino acid	+	+	Limited
Hemolymph plus lactalbumin hydrosate	+	+	−
Hemolymph plus yeast extract	+	+	−
Hemolymph plus heat killed *A. nematophilus*	+	+	−
Hemolymph plus egg albumin	−	−	−

TABLE 126

Mortality of *Galleria mellonella* Larvae when Injected with Culture Supernatent Supporting the Growth of *Neoaplectana carpocapsae* and *N. glaseri*

Supernatant from nematode species and strain	No. insects injected	No. insects died
Czechoslovakian strain of *N. carpocapsae*	15	11
DD-136 strain of *N. carpocapsae*	15	13
N. glaseri	10	3
Control media	15	0

4. Culture

a. In Vivo Methods

The first method of culturing *N. carpocapsae* in the laboratory involved the use of insects, namely larvae of the greater wax moth, *Galleria mellonella* (L.). In their original report of the DD-136 strain of *N. carpocapsae*, Dutky and Hough[223] mentioned that they commonly obtained yields in excess of 100,000 infective stage nematodes per infected wax moth larvae. This method was elaborated further in a later publication[248] when the authors claimed that each insect larva produced up to 200,000 infective stage juveniles. A simpler modification of the above was reported by Poinar.[1] Infection is achieved by adding infective stage juveniles to a standard petri dish containing two

FIGURE 46 and 47.   (46) Larvae of *Galleria mellonella* placed on a wet filter pad for extraction of *Neoaplectana carpocapsae* infective juveniles. (47) Strands of *Neoaplectana carpocapsae* infective stage juveniles (arrows) leaving a dead *Galleria mellonella* larva.

filter papers and from 20 to 30 last instar wax moth larvae (Figure 46). Actually, any large bodied lepidopteran would make a suitable rearing host, but *Galleria mellonella* is used because of its ease in laboratory rearing. After death, the infected insects can be kept in the infection chamber for a week, then removed and placed in a nematode collecting dish, similar to the one described by White.[249] The collecting dish consists of a filter paper on the lid of a small petri dish placed open side down on the bottom of a deep petri dish. Water is added to a depth of one half the height of the small petri dish so that the edges of the filter paper are in contact with the water. The parasitized insects are placed in the middle portion of the filter paper, and the infective stage juveniles emerge from the cadavers and migrate down the wet filter paper into the water (Figure 47). There they can be collected, concentrated in a separatory funnel, and stored in small flasks.

Carne and Reed[250] described another kind of apparatus for harvesting nematodes emerging from insect cadavers. Their system consisted of a series of interconnected funnels containing water at a high, constant level. Infected hosts were placed on a nematode-permeable diaphragm that rested in the funnel mouth. The stem of each funnel was connected through a glass T-junction to an outlet that could be periodically opened to harvest the nematodes as they left the host and passed through the diaphragm.

The following method can be used for rearing larvae of the greater wax moth at room temperature on a simple artificial diet.

Rearing medium —
1. Mix together 100 cc water, 100 cc honey, 100 cc glycerine, and 5 cc Deca Vi-Sol (or equivalent vitamins).
2. Pour the liquid mixture into 1200 cc dry Pablum or Gerber's mixed cereal.
3. Mix until homogenous and place in a ½-gal mason jar with a screen top. (A circle may be cut from a regular top and a piece of ordinary window screen cut to fit.)
4. Place 200 to 300 freshly collected eggs on the medium and incubate at 30°C, or at room temperature (30°C is preferable).

The larvae will reach the last instar in 4 to 5 weeks at 30°C and take about a week or more at room temperature. The culture may be stored at 10°C for up to 3 months, after which it should be returned to rearing temperature and the remaining larvae allowed to pupate. Emerging adults are collected by anesthetizing with CO_2 or are chilled in a refrigerator and transferred to a clean gallon mason jar with a screen top. New adults are added to the jar as they emerge. It is not necessary to provide the adults with food or water. A piece of accordion-pleated wax paper provides an oviposition site, and unfolding the wax paper causes the eggs to fall off.

Occasionally, it is necessary to initiate an infection using a small number of infective stages, especially when only a single host has been found infected in the field. Under these circumstances, the technique of forced feeding or "per os" injection can be employed. The first step is to fit a 1-cc tuberculin syringe with a glass tipped needle. The preparation of this needle and the method of injection is described below.

Glass-tipped feeding needle — The feeding needle is prepared by drawing out a fine capillary tube so it will fit into the mouth of the test insect. Cut the capillary at the small end and round this in a light flame, being careful not to seal the opening. The large end is sealed to the base of a 27-gauge needle with a drop of Duco® cement.

Method — The nematodes are drawn into a sterile syringe fitted with a feeding needle. The syringe is mounted on a microinjector and the delivery calculated. The tip of the needle is teased into the mouth of a test insect and the desired amount of inoculum injected. Care must be taken not to puncture the esophagus, as this will allow normal gut bacteria to enter the hemocoel and cause a fatal septicemia. After injection, the insect is reared on normal food. Several insects should be used for each test, and a control should be run using sterile saline or sterile distilled water.

If the "per os" injection technique is not successful, or if only a dozen or so infective stage juveniles are present, then the blood drop technique can be used for building up nematode populations. The technique is to externally sterilize the infective stage juveniles by placing them for 2 min in an aqueous solution of 0.1% Hyamine 10-X® (Rohm and Haas Company, Washington Square, Philadelphia, Pa.). Then transfer the surface-sterilized nematodes to a hanging drop of insect blood on a cover slip. Invert the cover slip and the infective stages should develop in the presence of their associated bacteria. (This is also a very successful method for isolating the symbiotic bacteria from individual infective stages.) As the nematodes develop, additional drops of blood can be added to the original drop, depending on the size of the population desired. Or the mature, mated females could be transferred to separate blood drops and allowed to oviposit. Although this is really an in vitro method, it is discussed here since living insect larvae are necessary for its use.

b. In Vitro Methods

Bixenic cultivation — One useful feature of neoaplectanid nematodes is that they can be grown on a variety of artificial media. Probably *N. carpocapsae* has been experimented with more than another other member of the genus in regards this aspect. Since the associated bacterium *Achromobacter nematophilus* supplies nourishment to the nematode, it is convenient to develop both organisms together. The bacterium does not supply all the necessary nutrients, however, as can be demonstrated by attempting to grow the nematode-bacterium complex on sterile agar plates. The present author has only been able to obtain up to 2½ generations of *N. carpocapsae* on any kind of simple bacteriological agar. The most successful tests were conducted on brain-heart infusion agar and resulted in 2½ generations, but after that the nematodes just stopped reproducing, even when transferred to new plates seeded with *A. nematophilus*. It was obvious that some limiting factor had halted reproduction and forced the remaining juveniles into the infective stage. If, however, a richer medium is used, then continuous culture can be achieved. Perhaps the most simple medium used thus far has dog food as its base, as described by House et al.[251] Various modifications of their basic preparation are now in existence (Figure 48). This culture method was used by Nutrilite Products to mass produce the DD-136 strain. One technician could handle about 200 petri dishes of dog food medium a day and could collect about 100 million nematodes each week. The cost of production (mainly labor) was about one dollar for one million nematodes.

Bedding[243] tested several solid media and various inert materials to increase the yield of neoaplectanid species. High productivity of the *Agriotos* strain was achieved by using coarse Aspen wood wool coated with homogenized chicken heart with 30% water added. The wood wool was soaked in homogenate, autoclaved, and packed into containers. Yields of infective stage nematodes reached 100 million per 2 *l* of container, and the total cost of rearing was about two cents per million nematodes. In this system, only *A. nematophilus* was associated with the nematode on the media. Bedding also remarked that "Monoexincity is essential since almost all foreign organisms rapidly render the substrate unsuitable."

Dutky et al.[248] kept stock cultures of the DD-136 strain of *N. carpocapsae,* together with *A. nematophilus,* on slants of peptone glucose agar containing a piece of cooked kidney and on a medium consisting of reconstituted dried whole egg solids and water. Additional media studies conducted by Nutrilite Corporation showed that good nematode development was achieved on cultures using wheat, wheat bran, corn, corn gluten oatmeal, whole fish, and several fly media as the basic constituent.[380]

Actually, just about anything high in protein can be used for culturing *N. carpocapsae* as long as two criteria are met. The first is the need for the continual association of *A. nematophilus* with the nematode. The second is the necessity of keeping foreign microorganisms (especially bacteria) out of the culture system. The latter is the most frequent cause of failure of artificial cultivation of this nematode. El-Sherif[381] was

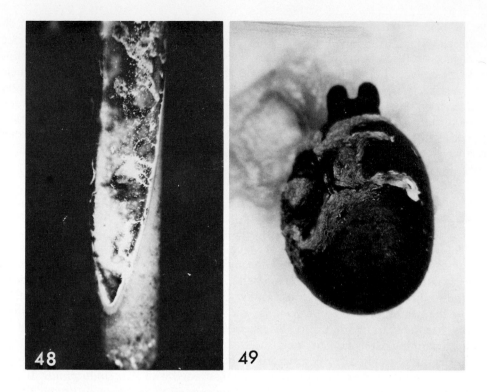

FIGURE 48 and 49.   (48) *Neoaplectana carpocapsae* developing in a tube of dog food agar. (49) Infective juveniles of *Neoaplectana carpocapsae* emerging from a pupa of the tse-tse fly, *Glossina morsitans.*

even successful in developing *N. carpocapsae* on an autoclaved medium composed mainly of chicken intestines, and Veremchuk[376] cultured *N. carpocapsae* on clotted cattle blood.

Axenic cultivation — Weiser[163] first axenized the Czechoslovakian strain of *N. carpocapsae* and maintained it on kidney-liver slants similar to the method used by Glaser[208] for growth of *N. glaseri*. Poinar and Thomas[205] first axenically cultured the DD-136 strain on an autoclaved medium consisting of 10 g ground beef kidney, 10 g ground beef liver, 0.5 g sodium chloride, 100 m*l* water, and 0.5 g agar. This medium was similar to the one used by Glaser et al.[219] for *N. chresima*.

Hansen and Cryan[252] were the first to culture *N. carpocapsae* in liquid media similar to methods developed for free-living nematodes. They produced continuous cultures in vessels in which the liquid was distributed as a film over glass wool which lined the inside surface of the vessel. The basic medium was a chemically defined mixture prepared by Grand Island Biological Company, Grand Island, New York. It was supplemented with 10 g/m*l* of heated liver extract. These continuous cultures (20°C) were harvested and the medium replenished biweekly. The harvested population contained 3000 nematodes per m*l* and comprised 52% infective stages, 6% juveniles, and 42% adults in equal sex ratio. With a supplement of Ficoll-activated growth factor at 250 μg/m*l*, a continuous culture was maintained for 18 months and comprised 50% infective stages and 10% adults.

5. Host Range

More experimental infections have been attempted with *N. carpocapsae* than with any other entomogenous nematode. The fact that neoaplectanid nematodes carry around a specific bacterium that overwhelms the host and establishes ideal conditions for nematode reproduction results in a wide host range for this group. Since the hosts are invariably dead when most nematode development occurs, problems that other insect nematodes face such as host reactions, adaptation to host cycles, etc. do not play a role in neoaplectanid infections.

It is interesting to note that *N. carpocapsae* will infect a number of medically important insects, i.e., mosquitoes, blackflies, tsetse flies, and others. Unfortunately, ecological and behavioral factors limit natural infections with the majority of these experimentally infected insects.

Niklas[372,373] accumulated host records for *N. carpocapsae* and Table 127 cites a complete, up-to-date list of both experimental and natural hosts of this nematode. Although adult honey bees are susceptible to infection by *N. carpocapsae,* larvae of the coccinellid (*Hippodamia convergens*) and lacewings (*Hemerobius* and *Chrysopa* spp.) are fortunately resistant in laboratory tests conducted by Berlowitz.[382]

6. Field Trials

N. carpocapsae has been used against a variety of insects under various field conditions. These trials are summarized in Table 128. In all of these cases, the nematodes were apparently cultured in larvae of *Galleria mellonella,* and the infective stage juveniles were held in water under refrigeration.

Storage — The infective stage juveniles of *N. carpocapsae* will remain infective for long periods in water at low temperature (6 to 9°C) (Figure 50). If bacterial contamination is a problem, then a 0.001% aqueous solution of formalin can be used. Other methods of storage (lyophilization, freezing, drying) are still at the experimental stage.

Shipping — An efficient, inexpensive method of shipping *N. carpocapsae* is difficult to achieve. The most important aspect involved in shipping nematodes is to be able to provide enough air to keep them alive en route. This can be achieved by plugging the containers with cotton and avoiding the use of water as a shipping medium. A good method might be to place the infective stages on some inert substances (charcoal, wood chips) that can be kept moist. However, the nematodes should be able to be removed easily from the substrate in water when ready for application.

Application — The infective stage juveniles have been applied to the substrate in a variety of ways. The simplest method has been to dilute the nematode suspension with water and apply the mixture directly to the soil or plant surface with a watering can, pipette, or sprayer. Various antidesiccating agents can be added to the suspension to prolong the life of the infective stages on plants.

Recently, Bedding[243] discussed the advantages of enclosing the nematodes in a paraffin-oil mixture to prevent desiccation after application. Apparently, there is much promise in the use of additives, and these will certainly increase the possible number of habitats in which these nematodes can be used. Placing laboratory-infected *Galleria mellonella* larvae in soil and allowing the infective stage nematodes to emerge naturally is another method of application.

Safety — Some tests have been conducted to determine the safety of *N. carpocapsae.* In 1969, Nutrilite Corporation fed five million nematodes to each of ten 3-month-old albino rats over a 5-day period. Fecal samples were examined, and weight gain was recorded for the experimental and control animals for 20 days after feeding was terminated. There was no clinical evidence of any physiological disturbance in the experimental group during the study period. Nematodes were recovered from feces of experimental animals during the feeding but not in the subsequent period. Further discussion of this subject is given in the section on environmental impact.

M. *Neoaplectana bothynoderi* Kirjanova and Putschkova 1955

The present species was found in the U.S.S.R. by Kirjanova and Putschkova[296] parasitizing larvae of the beet weevil *Bothynoderes punctiventris*. To the author's knowledge, it has never been reisolated, and the only information on this species is in the original publication.

TABLE 127

Host Range of *Neoaplectana carpocapsae* Weiser

Order	Family	Host	Stage[a]	Infection[b]	Ref.
Coleoptera	Anobiidae	*Lasioderma serricorne* (Fab.)	L	E	223
	Bostrichidae	*Dinoderus minutus* (Fab.)	A	E	380
	Bruchidae	*Acanthoscelides obtectus* Say	A	E	380
	Carabidae	*Zabrus tenebrioides* Goeze	L	N	253
	Cerambycidae	*Aubacophora foveicollis* Lucus	L	E	254
		Chlorida festiva L.	A	E	380
		Eburia octomaculata Chev.	A	E	380
		Monochamus scutellatus (Say)	L, P	E	254
	Chrysomelidae	*Cassida rubiginosa* Muller	L, A	E	255
		Chaetocnema confinis Crotch	L	E	256
		Diabrotica balteata Le Conte	L	E	256
		D. unipunctata howardi (Barber)	L, A	E	143
		Leptinotarsa decemlineata (Say)	L, A	E	259
		Phyllotreta sp.	L, P	E	380
		Plagiodera versicolora (Laich.)	L, A	E	258
		Psylliodes chrysocephala (L.)	L	E	259
		Systena spp.	L	E	256
	Cleridae	*Madoniella pici* Ep.	A	E	380
	Coccinellidae	*Coleomegylla* sp.	A	E	380
		Cycloneda sanguinea L.	A	E	380
		Epilachna varivestis Mulsant	L, P, A	E	258
		Epilachna vigintiocto-maculata (Motsch.)	L, A	E	255
		Propilia 14 punctata L.	L	E	380
	Curculionidae	*Aegorhinus phaleratus* Erich.	L, P	E	258
		Anthonomus grandis Boh.	A	E	258
		A. vestitus Boh.	A	E	260
		Baris caeralesceus Scop.	L	E	380
		Brachycerus undatus Fab.	L	E	380
		Ceutorhynchus assimilis Payk.	L	E	259
		C. napae Gyll.	L	E	259
		C. picitarsis Gyll.	L	E	259
		C. pleurostigma Marsh.	L	E	259
		C. quadridens Panz.	L	E	259
		Cosmopolites sordidus Germ.	A	E	380
		Curculio caryae (Horn)	L	E	261
		Diaprepes abbreviata L.	A	E	380
		D. famelicus Oliv.	A	E	380
		D. marginatus Oliv.	A	E	380
		Graphognathus leucoloma Buch.	L	E	258
		Graphognathus peregrinus (Buch.)	L	E	262
		Heilipus latro Gyll.	L	E	380
		Hylobius radicis Buch.	L	E	234
		Hypera postica (Gyll.)	L, P	E	258
		Lasioderma serricorne (Fab.)	L, A	E	223
		Lissorhoptrus oryzophilus (Kusch.)	L	E	143
		Listroderus costirostris obliquus (Klug)	L	E	258
		Metamesius hemipterus L.	A	E	380
		Otiorrhynchus sulcatus Fab.	L	E	380
		Pantomorus sp.	L	E	258
		Pissodes notatus Fab.	L, P	E	380

TABLE 127 (continued)

Host Range of *Neoaplectana carpocapsae* Weiser

Order	Family	Host	Stage[a]	Infection[b]	Ref.
	Dermestidae	*Dermestes vulpinus* Fab.	L	E	258
	Elateridae	*Agriotes lineatus* L.	L, P, A	N	189
		Conoderus falli Lane	L	E	256
		Eburia octomaculata Chevr.	A	E	380
		Ectinus dahuricus	L	E	255
		Selatosomus reichardti	L	E	255
	Lystidae	*Lyctus* sp.	A	E	258
	Nitidulidae	*Carpophilus hemipterus* (L.)	L, A	E	263
		C. humeralis (Fab.)	L, A	E	263
		C. multilatus Erich.	L, A	E	263
		C. obsoletus Erich.	L, A	E	263
		Haptoncus luteolus (Erich.)	L, A	E	263
		Meligethes sp.	L	E	259
		Stelidota geminata (Say)	L, A	E	263
	Passalidae	*Passalus unicornis* Serv.	A	E	380
		Paxillus puncticollis St. F. & S.	A	E	380
	Scarabaeidae	*Ectinohoplia rufipes* Motsch.	L, A	E	255
		Hylamorpha elegans (Burm.)	L	E	258
		Macraspis tristis (Cast.)	A	E	380
		Melolontha hippocastani Fab.	L	N	264
		M. melolontha (L.)	L	N	264
		Othnonius batesi Olliff	L	E	264
		Plectris aliena Chapin	L	E	256
		Premnotrypes vorax (Hust.)	L	E	265
		Sericesthis germinata Bois.	L	E	244
	Scolytidae	*Dendroctonus frontalis* Zimm.	L, A	E	266
		Hexacolus guyanensis Schedl.	L, P, A	E	380
		Scolytus scolytus Fab.	L, A	E	267
	Tenebrionidae	*Tenebrio molitor* L.	L	E	258
		Tribolium castaneum Hbst.	L, A	E	268
		Zophobas atratus Fab.	A	E	380
Diptera	Anthomyidae	*Hylemya brassicae* (Boreche)	L	E	238
		H. floralis (Fallen)	L	E	255
		H. platura (Mg.)	L	E	269
	Cecidomyiidae	*Dasyneura brassicae* Winn.	L	E	259
	Culicidae	*Aedes aegypti* (L.)	L	E	231
		A. sierrensis (Ludlow)	L	E	143
		Culex pipiens L.	L	E	230
		Culiseta inornata (Will.)	L	E	270
	Muscidae	*Glossina morsitans* Westw.	P, A	E	233
		Musca domestica L.	L, P	E	258
	Simuliidae	*Simulium vittatum* Zett.	L	E	270
	Tachinidae	*Metagonistylum minense* Townsend	A	E	380
	Tephritidae	*Anastrepha ludens* (Loew)	L, A	E	258
	Tipulidae	*Tipula paludosa* (Mg.)	L	E	295
	Trypetidae	*Rhagolites pomonella* (L.)	P	N	143
Heteroptera	Lygaeidae	*Blissus leucopterus insularis* (Barb)	N, A	E	258
	Pentatomidae	*Nezara viridula* (L.)	A	E	380
	Pyrrhocoridae	*Dysdercus cingulatus* Fab.	A	E	254
		D. peruvianus (Guerin)	A	E	260
		Pyrrhocorus sp.	A	E	380
	Reduviidae	*Rhodnius prolixus* Stal.	N, A	E	271
		Triatoma infestans King	N, A	E	271
	Tingidae	*Corythaica cyathicollis* (Costa)	A	E	380
Homoptera	Aphididae	*Brevicoryne brassicae* (L.)	J, A	E	258
	Coreidae	*Phthia picta* Drury	L, A	E	380
	Margarodidae	*Margarodes vitium* Giard	A	E	258
	Pseudococcidae	*Planococcus citri* (Risso)	J, A	E	258

TABLE 127 (continued)

Host Range of *Neoaplectana carpocapsae* Weiser

Order	Family	Host	Stage[a]	Infection[b]	Ref.
Hymenop-tera	Apidae	*Apis mellifera* L.	L	E	258
		A. mellifera L.	A	E	272
	Diprionidae	*Diprion similis* (Hartig)	L	E	234
		Neodiprion americanus pratti (Dyar)	L	E	253
		N. lecontei (Fitch)	L	E	234
		N. sertifer (Geoffr.)	L	E	163
	Formicidae	*Acromyrmex octospinosus* (Reich)	L	E	273
		Camponotus sp.	A, L	E	380
		Solenopsis geminata (Fab.)	A	E	380
	Scoliidae	*Campsomeris dorsata*	A	E	380
	Sphecidae	*Sphex calignosus* Erick.	A	E	380
	Tenthredinidae	*Athalia proxima* Klug	L	E	254
		Athalia rosae L.	L	E	380
		Caliroa aethiops Fab.	L	E	258
		Pikonema alaskensis (Roh.)	L	E	234
		Pristiphora californica (Marl.)	L	E	258
		P. erichsonii (Htg.)	L, P	E	234
	Vespidae	*Polystes* sp.	L, A	E	380
		Vespula pensylvanica (Sauss.)	A	E	274
		V. rufa atropilosa (Sladen)	A	E	274
	Xylocopida	*Xylocopa mordax* L.	A	E	380
Isoptera	Termitidae	*Coptotermes formosanus* Shiraki	A	E	275
		Nasutitermes costalis Holm.	A	E	380
		Termes sp.	A	E	258
Lepidop-tera	Aegeriidae	*Vitacea polistiformis* (Har.)	L	N	227
	Arctiidae	*Diacrisia obliqua* Walker	L	E	254
		Estigeme acrea (Drury)	L	E	258
		Hyphantria cunea (Drury)	L	E	258
	Bombycidae	*Bombyx mori* (L.)	L	E	260
	Dioptidae	*Phryganidia californica* Pack.	A	E	232
	Galleriidae	*Galleria mellonella* (L.)	L, P	E	223
		G. mellonella (L.)	A	E	232
		Gnoremoschema opercullella (Zeller)	L	E	258
	Gelechiidae	*Anarsia lineatella* Zeller	L, P, A	E	276
		Pectinophora gossypiella (Saund.)	L	E	253
	Geometridae	*Alsophila pometaria* (Harris)	L	E	277
		Hibernia defoliaria (L.)	L	E	163
		Operophthera brumata (L.)	L	E	277
	Hesperiidae	*Urbanus proteus*	L	E	380
	Hepialidae	*Maculella noctuides* (Pfitz.)	L	E	258
		Oncopera fasciculata (Walk.)	L	E	244
	Hyponomeutidae	*Acrolepia assectella* Zeller	L, A	E	380
		Hyponomeuta malinellus Zeller	L	E	163
		Plutella maculipennis Curt.	L	E	380
	Lasiocampidae	*Dendrolimus pini* (L.)	L	E	163
		Malacosoma americanum (Fab.)	L	E	258
		M. californicum (Pack.)	L, P, A	E	276
		M. neustria (L.)	L	E	163
	Limacodidae	*Parasa variabilis* Butlr.	L	E	380
	Lymantriidae	*Porthetria dispar* (L.)	L	E	163
	Lyonetiidae	*Hieroxestis subcervinella* Meyr	L	E	380
	Noctuidae	*Agrotis ipsilon* (Huf.)	L	E	278
		A. segetum (Schiff)	L	E	268
		A. subterranea Fab.	L	E	380
		Alabama argillacea (Hbn.)	L	E	258

TABLE 127 (continued)

Host Range of *Neoaplectana carpocapsae* Weiser

Order	Family	Host	Stage[a]	Infection[b]	Ref.
		Barathra brassicae L.	L	E	268
		Cirphis compta Mo.	L	E	279
		Earias sp.	L	E	380
		Euxoa messoria (Harris)	L	E	242
		Heliothis armigera (Hbn.)	L	E	258
		H. virescens (Fab.)	L	E	260
		H. zea (Boddie)	L	E	280
		Hydraecia xanthenes Germ.	L	E	380
		Laphygma frugiperda (Smith)	L	E	258
		Mamestra oleracea L.	L	E	380
		Mocis punctularis Fab.	L	E	380
		Peridroma saucia L.	L	E	278
		Plusia gamma L.	L	E	278
		Prodenia eridania (Cramer)	L	E	380
		Pseudaletia separata Walker	L	E	279
		P. unipuncta (Haw.)	L	E	258
		Pyrausta nubilalis (Hbn.)	L	E	238
		Scotogramma trifoli Rott.	L	E	268
		Spodoptera frugiperda (Smith)	L	E	375
		Spodoptera littoralis Boisd	L	E	380
		S. litura Fab.	L	E	254
		Trichoplusia ni (Hbn.)	L	E	258
	Notodontidae	*Symmerista albifrons* (A. and S.)	L	E	234
	Olethreutidae	*Epinotia aporema* Wals.	L	E	258
		Grapholitha molesta (Busk.)	L, P, A	E	276
		Pseudexentera mali (Freeman)	L	E	277
		Rhyacionia buoliana (Schiff.)	L	E	234
		R. frustrana (Comstock)	L	E	258
	Papilionidae	*Papilio demodocus* Esper.	L	E	380
		P. epiphorbas Boisd	L	E	380
		P. machaon L.	L	E	380
	Pieridae	*Pieris brassicae* (L.)	L	E	268
		P. rapae (L.)	L	E	238
	Plutellidae	*Acrolepia assectella* (Zeller)	L	E	278
	Pyralidae	*Achroia grisella* (Fab.)	L	E	260
		Chilo suppressalis (Wlk.)	L, P	E	281
		Diaphania hyalinata L.	L	E	380
		Diatraea saccharalis (Fab.)	L	E	258
		Elasmopalpus lignosellus Zeller	L	E	380
		Ephestia elutella (Hbn.)	L	E	223
		E. kuehniella (Hbn.)	L	E	258
		Hypsipila grandella (Zeller)	L	E	380
		Leucinodes orbonalis (Guen)	L	E	254
		Mescinia peruella Schaus	L	E	260
		Mesocondyla condordalis Hb.	L	E	380
		Ostrinia nubilalis (Hbn.)	L	E	223
		Paramyelois transitella (Walk.)	L, P, A	E	276
		Platyptilia carduidactyla (Riley)	L	E	282
		Plodia interpunctella (Hbn.)	L	E	258
		Pococera atramentalis (Led.)	L	E	260
		Pyrausta nubilalis (Hbn.)	L	E	283
		Scirpophaga nivella (Fab.)	L	E	283
		Sesamia inferens Walker	L, P	E	281
		Tryporyza incertulas (Walk.)	L	E	283
		T. incertulas (Walk.)	A	E	284
		Zinkenia fascialis (Cramer)	L	E	380

TABLE 127 (continued)

Host Range of *Neoaplectana carpocapsae* Weiser

Order	Family	Host	Stage[a]	Infection[b]	Ref.
	Saturniidae	*Antheraea pernyi* (Guerin)	L	E	163
	Sphingidae	*Herse convolvuli* L.	P	E	380
		Pseudosphinx tetrio (L.)	L	E	380
		Protoparce sexta Johan.	L	E	258
	Tortricidae	*Argyrotaenia citiana* (Fernald)	L	E	285
		Choristoneura fumiferana (Clemens)	L, P	E	234
		C. pinus Freeman	L, P	E	234
		C. rosaceana (Harris)	L, P, A	E	276
		Laspeyresia pomonella (L.)	L	N	222
		Paramyelois transitella (Walker)	L	E	286
		Yponeumata padella (M)	L, P	E	183
	Zygaenidae	*Harrisina americana* (Guerin)	L	E	258
Neuroptera	Chrysopidae	*Chrysopa collaris* Schn.	A	E	380
Odonata		*Anax junius* Drury	A	E	380
		Mesocondyla concordalis	A	E	380
		Turpilia rigulosa	A	E	380
Orthoptera	Arcididae	*Amphitornus coloradus* (Thomas)	A	E	258
		Bruneria brunnea (Thomas)	A	E	258
		Melanoplus bruneri Scudder	A	E	258
		M. mexicanus Saussure	A	E	258
		M. packardii Scudder	A	E	258
		Phlebestroma quadrimaculatum (Thomas)	A	E	258
	Blattidae	*Blatella germanica* (L.)	J, A	E	258
		Nauphoeta cinerea (L.)	J, A	E	258
		Periplaneta americana (L.)	A	E	258
	Tettigoniidae	*Anabrus simplex* Hald.	A	E	258
	Truxalidae	*Aulocara elliotti* (Thomas)	A	E	258
	Gylloidea	*Acheta domestica* L.	A	E	183
		A. assimilis Fab.	A	E	380
		Gryllotalpa gryllotalpa (L.)	A	E	183
		Neocurtilla hexadactyla Perty	A	E	380

[a] L, larva; P, pupa; A, adult; J, juvenile.
[b] N, natural infection; E, experimental infection.

1. Description[296]

Adults — Females are variable in size. The range of the "a" value varies from 5.6 to 23.4; that of "b" varies from 8.2 to 11.7 and "c" from 17.5 to 31.3. The eggs range from 35 to 45 μm in diameter, and the juveniles hatch in the uterus of the mother nematode. The anterior end bears 12 small labial papillae. Small tubercle-like teeth are found on either side of the buccal cavity. The pharynx has a basal valvated bulb. The tail ends in a hair-like projection. In males, the terminal portions of the spicules and gubernaculum bear denticles. The tail terminus bears a distinct filament. Quantitative values of the adult nematodes are presented in Tables 129 and 130.

Modified description — During a visit to Leningrad, the present author examined the type material of *N. bothynoderi* in the laboratory of Dr. Kirjanova, and the following comments are based on that examination. The presence of denticles on the tip of the spicules and gubernaculum could not be verified; however, the tip of the spicules was blunter than those of *N. carpocapsae,* and the male tail projection was longer than the genital papillae (22 plus 1), but in all other aspects the nematodes resembled *N. carpocapsae.*

TABLE 128

Field Trials of *Neoaplectana carpocapsae* Against Insect Populations

Host	Site	Dosage	Results	Location	Ref.
Colorado potato beetle *Leptinotarsa decemlineata*	500 potato plants	20,000 nemas per plant	At least 14% reduction over control plot	Canada	257
Dysdercus peruvianus (Hemiptera)	Cotton plants	48,000 to 144,000 per plant	22 to 36% mortality	Peru	260
Tobacco budworm *Heliothis virescens*	Tobacco leaves	5×10^6 nemas per plant	80 to 85% reduction under very moist conditions	Eastern U.S.	287
Codling moth *Laspeyresia pomonella*	Trunks and branches of apple trees	—	60% and higher mortality	Eastern U.S.	237
Artichoke plume moth *Platyptilia cardiudactyla*	Head of 202 artichoke plants	2,000 and 4,000 nemas per plant	No significant reduction	California	282
Colorado potato beetle *Leptimotarsa decemlineata*	640 potato plants	From 5×10^4 to 40×10^4 nemas per plant	Small reductions; no significant difference	Ontario, Canada	288
Cabbage root maggot *Hylemya brassicae*	Cabbage, radish, and rutabaga plants	1.0×10^5 to 5.1×10^5 nemas per plants; cadaver burial	Reduction in plant damage over control	Ontario, Canada	238
European corn borer *Ostrinia nubilalis*	Leaf sheath of corn plants	10,000 nemas per leaf sheath	Reduction of cob damage in treated plants	Ontario, Canada	238
Imported cabbage worm *Pieris rapae*	Cabbage head	3.4×10^5 to 1.5×10^6 nemas per plant	No significant reduction over controls	Ontario, Canada	238
Root collar weevil *Hylobius radicis*	Base of ten pine trees	10,000 infective nemas per tree	Infection obtained	Minnesota	234
Corn earworm *Helicoverpa zea*	Corn silk, four to seven rows of plants	2,000 to 6,000 nemas per ear	Mortality high, damage to corn not prevented	California	280
Corn earworm *Helicoverpa zea*	50 corn ears	4,000 to 9,000 nemas per ear	From 19 to 59% mortality	Missouri	240
Imported cabbage worm *Pieris rapae*	Cabbage heads	1,840 to 3,120 nemas per ml spray	Significant mortality	Canada	289

TABLE 128 (continued)

Field Trials of *Neoaplectana carpocapsae* Against Insect Populations

Host	Site	Dosage	Results	Location	Ref.
Winter moth *Operophtera brumata*	Apple trees (10 ft tall)	11.25×10^6 nemas per tree	No significant reduction	Canada	277
Pruinose scarab *Sericesthis geminata* and dark soil scarab *Othnonius batesi*	Turf plot	35 to 70×10^3 nemas per ft²	No nema-infected larvae found	Australia	244
Banded cucumber beetle *Diabrotica balteata*	Plots in sweet potato field	0.54 to 1.09×10^9 nemas per acre	Reduction of damage but protection inadequate	Southern U. S.	256
Pale apple leaf roller *Pseudexentera mali*	Soil beneath apple trees	3.4×10^4 to 1.6×10^5 nemas per ft²	Reduced larvae to one third control	Canada	290
Winter moth *Operophtera brumata*	Soil beneath apple trees	3.4×10^4 to 1.6×10^5 nemas per ft²	Reduced survival of cocoons to 12%	Canada	290
Cutworm *Pseudaletia separata*	Rice field	1×10^6 nemas to a 3.0×6.5 m² plot	Significant reduction of larvae	India	279
Paddy cutworm *Cirphis compta*	Rice field	1×10^6 nemas to a 3.0×6.5 m² plot	Significant reduction of larvae	India	279
Nantucket pine tip moth *Rhyacionia frustrana*	Pine branches	4×10^3 nemas per ml	5 to 15% reduction of moth populations	South Carolina	291
Southern pine beetle *Dendroctonus frontalis*	Pine bark	740 nemas per ft²	40 to 50% mortality of brood and adults	North Carolina	266
Wireworm *Agriotes* sp. Pear aphids Leaf beetles Ladybird larvae Root fly larvae	Various	$2.5\text{-}3 \times 103$ per ml	60% mortality of wireworms; 70% mortality pear aphids; 33 to 60% mortality leaf beetles; 30% mortality ladybird larvae; 5 to 10% mortality root fly larvae	U.S.S.R.	255
Rice stem borer *Chilo* sp.	Rice paddy	8 spray treatments	Mortality of *Chilo* larvae noted	India	292

TABLE 128 (continued)

Field Trials of *Neoaplectana carpocapsae* Against Insect Populations

Host	Site	Dosage	Results	Location	Ref.
Yellow moth borer *Tryporyza incertulas*	Rice paddy 1. Microplots 2. Field plots	Spray treatment	1. Some borers infected 2. No difference between treated and control plots	India	284
White-fringed beetle *Graphognathus peregrinus*	Grassland	4.3×10^5 nemas per m² to 12 plots	38% reduction in larvae	Louisiana	262
Formosan termite *Coptotermes formosanus*	Termite colonies	Infected termites	Infection obtained	Hawaii	275
Onion borer *Acrolepia assectella*	Soil	Not stated	High mortality at pupations	France	278
Hylemya spp.	Tobacco	38,200 nemas per plant	Control by nemas equal to that of Diazinon	Ontario	293
Pecan weevil *Curculio caryae*	Pecan tree	703,000 nemas per inch drain tiles	67% mortality of weevil larvae	Georgia, U.S.	261
Spodoptera frugiperda	Field of maize	4,000 nemas per maize plant	50 to 60% decrease in insect larvae	Colombia, S.A.	294
Rice stem borer	Rice stubble	Diluted spray suspension	Complete mortality achieved	Japan	281
Navel orangeworm *Paramyelois transitella*	Almond orchard (four trees)	46 to 1,500 nemas per almond	24 to 100% mortality in almonds	California	286
Wireworms *Agriotes lineatus Selatosomus aeneus*	Soil	1.5 million nemas per m²	67 to 76%	U.S.S.R.	379

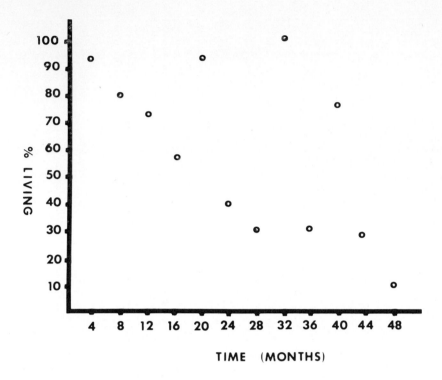

FIGURE 50.   Longevity of *Neoaplectana carpocapsae* infective stages held in water at 8 to 10°C.

TABLE 129

Measurements of *Neoaplectana bothynoderi* Females[296]

Character	Value
Total length (mm)	1.2—2.2
Greatest width	100—253
Length stoma	8
Width stoma	12
Length head to base of esophagus	188
Length head to nerve ring	130
% vulva	48—58
Width at anus	132

Diagnosis — Again, it is not known whether Kirjanova and Putschkova described the first or second generation adults, but from the lengths presented, it was probably the second generation. The only characters separating this species from *N. carpocapsae* are the blunt tip of the spicules and the pointed tail projection in both sexes, both of which are doubtful specific characters.

2. Bionomics

Life cycle — Kirjanova and Putschkova[296] reported that it took one nematode generation only 4 days to develop. They cultured several generations of *N. bothynoderi* on a small piece of tissue from the host's body.

3. Bacterial Associates

The fact that beetle larvae infected with *N. bothynoderi* remained perfectly intact

TABLE 130

Measurements of *Neoaplectana bothynoderi* Males[296]

Character	Value
Total length (mm)	0.74—0.94
Greatest width	104—107
Length stoma	9
Length head to base of esophagus	150
Length head to nerve ring	110
Length tail	60
Width at anus	65
Length spicule	64—65
Width spicule	12—15
Length gubernaculum	35—40
Width gubernaculum	6—10

after death suggests the presence of a symbiotic bacterium, although the authors did not comment further on this point. They did speculate that the nematodes secrete special substances which prevent decomposition of the host cadaver. It is now known that the associated bacterium of *N. carpocapsae* (e.g., *Achromobacter nematophilus*) somehow "preserves" the host for nematode development. Kirjanova and Putschkova[296] also mention that the host cadaver becomes pinkish-yellow and soft to the touch. This is characteristic of the action of an associated bacterium.

4. Host Range and Distribution

The type and only host mentioned for *N. bothynoderi* was the beet weevil *Bothynoderes punctiventris* Germ. Kirjanova and Putschkova[296] mention that the nematode is highly injurious to the larvae and would be easily cultured under laboratory conditions. It was collected in the Somenove district of the Poltava region, and the rate of infection in a field population of the insects reached 7.5%. Infections were found in the field only during July and August, and the rate of infection differed greatly from field to field. In an area near a forest belt, no infected weevil larvae were found, while in a field close to a dam, there were 16.4 infected insects per square meter.

N. *Neoaplectana melolonthae* Weiser 1958

This nematode was found in grubs of *Melolontha melolontha* and *M. hippocastani* L. in the region of the lower Morava river in Czechoslovakia.

1. Description[297,163]

Adults — Females have three pronounced lips. They contain 60 to 65 eggs which hatch in the uterus. The tail has a drawn-out pointed tip reaching 180 μm in length. Eggs are 60 × 40 to 45 μm in diameter. Quantitative measurements are given in Table 131. In males, the testis is outstretched. The male tail is drawn out into a long spike. There is one preanal papillae and 7 to 9 postanal papillae. Quantitative measurements are given in Table 132.

Infective stage juveniles — This stage is 580 μm long and 37 μm wide. The tail is 72 μm long.

Diagnosis — The shape of the tail of the male and female, the shape of the spicules and gubernaculum of the male, the enlarged corpus of the pharynx, the broad stoma, the lack of modified valves in the basal bulb, and the arrangement of the genital papillae all indicate that this nematode does not belong in the genus *Neoaplectana,* but somewhere in the Diplogasteridae.

TABLE 131

Measurements of *Neoaplectana melolonthae* Females[163,297]

Character	Value
Total length (mm)	2.74 (2.0—3.0)
Greatest width	207
Length stoma	30
Width stoma	9
Length head to base of esophagus	208
Length head to nerve ring	143
% vulva	41
Length tail	500
Width at anus	27

TABLE 132

Measurements of *Neoaplectana melolonthae* Males[163,297]

Character	Value	
	$\overline{X}$	Range
Total length (mm)	1.02	1.0—2.0
Greatest width	66	60—80
Length stoma	24	
Width stoma	9	
Length head to base of esophagus	210	
Length head to nerve ring	132	
Length tail	48	
Width at anus	48	
Length spicule	58	
Length gubernaculum	25	

O. *Neoaplectana georgica* Kakulia and Veremchuk 1965

This species was recovered from larvae of the June beetle *Amphimallon solstitialis* (L) in the Soviet state of Georgia in April 1964.

1. Description [298]

Adults — In females, the cuticle is annulated and thick. There are six lips. The tail is pointed. The eggs are nearly round, 30 to 42 × 30 to 39 μm in diameter. Quantitative measurements are given in Table 133. Males have ten pair and a single preanal genital papilla: five pair preanal, four pair postanal, and one pair located lateral dorsal. Spicules are orange in color, and the tail has a terminal spike. Quantitative measurements are given in Table 134.

Infective stage juveniles (N = 10) — This stage is 795 (780 to 820) μm long and 34 (30 to 37) μm wide. a = 23.3 (22.1 to 24.1), b = 6.3 (6.1 to 6.5), c = 12.2 (11.7 to 12.5).

Diagnostic characters — In their differential diagnosis, Kakulia and Veremchuk[298] compare *N. georgica* only with *N. feltiae,* listing size, shape of spicules, host, and body ratios or differential characters. Knowing how variable all of these characters are does not leave much as distinctive characters for *N. georgica*. The length of the infective stage juveniles, a fairly stable character, is approximately equal in both *N. georgica* and *N. feltiae;* however, the EP/pharynx ratio of the adults does seem to be a consistent difference.

TABLE 133

Measurements of Females of *Neoaplectana georgica*[298]

Character	Value
Total length (mm)	1.18 (0.94—1.73)
Greatest width	75 (66—114)
Length stoma	8.5 (7.2—12)
Length head to base of esophagus	156 (156—200)
Length head to excretory pore	100
Length head to nerve ring	140
% vulva	54 (52—56)
Length tail	44 (32—74)
Width anus	38

TABLE 134

Measurements of Males of *Neoaplectana georgica*[298]

Character	Value
Total length (mm)	0.92 (0.79—1.10)
Greatest width	44 (36—48)
Length head to base of esophagus	158
Length head to excretory pore	98
Length head to nerve ring	142
Length reflextion of testis	200
Length tail	28 (24—37)
Width at anus	42
Length spicule	48 (42—53)
Length gubernaculum	30 (24—34)
Length tail spike	9 (5—13)

2. Bionomics

Life cycle — The nematode developed through one generation in 3 to 4 days at 20 to 27°C in larvae of *Galleria mellonella*.[298] Infectivity tests against beetles, cutworms, and other insects were positive.[299] *N. georgica* was also cultivated on week-old individuals of *Passer domesticus* L., as well as on Martin meat peptone agar and other artificial media. Three generations are produced in June beetle grubs, and the species is considered viviparous with each female producing about 150 juveniles. Cultures died out when the temperature was lowered to 11°C.

The chafer *A. solstitialis* is the natural host for this species, and to the author's knowledge, *N. georgica* has never been used as a biological control agent.

P. *Neoaplectana arenaria* Artyukhovsky 1967

This species was recovered from larvae and pupae of the May beetle (*Melolontha hippocastani* Fab.) in 1957 in the Usmanski forest in the Voronezk region of the U.S.S.R. To the author's knowledge, it has not been recovered again in the Soviet Union.

1. Description[300]

Adults — Females are variable in length. The lips are distinct. The valve in the basal bulb of the pharynx is distinct. The excretory pore is anterior to the basal bulb of the pharynx. Eggs are 32 to 48 × 32 to 56 μm. In males, the excretory opening is near the

TABLE 135

Measurements of Females of *Neoaplectana arenaria*[300]

Character	Value
Total length (mm)	2.17—4.00
Greatest width	149—241
Length head to base of esophagus	225—257
Length head to nerve ring	177
% vulva	53—54
Length tail	96—97
Width at anus	48—64

TABLE 136

Measurements of Males of *Neoaplectana arenaria*[300]

Character	Value
Total length (mm)	1.03—1.83
Greatest width	60—97
Length head to base of esophagus	209
Length head to nerve ring	144—145
Length tail	38—48
Width at anus	41—42
Length spicule	60—72
Length gubernaculum	35—44

level of nerve ring. The tail is without a terminal spine or projection and is smooth. Spicules are without a hook or notch. There are five pair of postanal papillae, three pair of preanal papillae, and a single large adanal papilla. Quantitative values of the adults are present in Tables 135 and 136.

Infective stage juveniles — The length of this stage is 1143 to 1300 μm; the greatest diameter is 33 to 42 μm.

Diagnostic characters — The length of the infective stage juveniles, the absence of a tail projection on the male, and lack of a pronounced arch on the shaft of the spicules separates *N. arenaria* from all previously described neoaplectanids except *N. glaseri*. Artyukhovsky[300] stated that *N. arenaria* is distinguished from *N. glaseri* by the absence of hooks at the end of the spicules. However, in his Figure 2 on page 96, Artyukhovsky[300] indicates a small notch on the ventral surface of the spicule tip. Such a small notch is also typical of some *N. glaseri* males. The present author discovered that the notch is not consistently well-formed on all males of *N. glaseri* and is barely perceptible in some. The above characters, and also the posterior position of the excretory pore in both sexes, lead the present author to consider *N. arenaria* as another population of *N. glaseri* Steiner.

2. Bionomics[300]

Life cycle — The generation time lasted 6 to 7 days, and there were approximately 1.5 more males than females in newly infected hosts. The eggs hatched in the female's body, and the developing juveniles eventually consumed her. Egg production varied from 14 to 16 to 42 to 46 per female, depending on the amount of food available.

Ecology — Both larvae and pupae of *M. hippocastani* were attacked by nematodes in the end of June, but only in sandy soil of a mixed forest of pine and oaks. The rate of infection in third stage host larvae was 2 to 3% in one year and 6% at a later date.

TABLE 137

Measurements of *Neoaplectana kirjanovae* Females[301]

Character (N = 15)	Value	
	X̄	Range
Total length (mm)	2.95	2.72—3.40
Length stoma	7	6—13
Length head to base of esophagus	149	134—163
Length head to excretory pore	30	
Length head to nerve ring	60	
% vulva	54	53—56
Length tail	21	14—27
Width at anus	22	

TABLE 138

Measurements of *Neoaplectana kirjanovae* Males[301]

Character (N = 15)	Value	
	X̄	Range
Total length (mm)	1.04	0.96—1.17
Greatest width	87	71—106
Length stoma	8	6—10
Length head to excretory pore	50	
Length head to base of esophagus	145	133—159
Length tail	17	13—27
Width at anus	48	41—60
Length spicule	66	58—74
Length gubernaculum	41	40—48

3. Bacterial Associates

Artyukhovsky[300] wrote that host larvae infected with *N. arenaria* were deformed and brown in color. Their body contents was viscous and contained many juvenile nematodes. This description indicates the presence of an associated bacterium. Japanese beetle grubs infected with *N. glaseri* turn a reddish-brown color, and their body contents break down into a viscous mass.

Q. *Neoaplectana kirjanovae* Veremchuk 1969

This species was collected in August 1962 in the Leningrad district from an elaterid pupa.

1. Description[301]

Adults — In females, the cuticle is thin, 1 μm thick. Eggs are 36 (36 to 37) μm in diameter. The tail tip has a blunt knob. Quantitative measurements are given in Table 137. In males, the tail is broad, the tip without a projection. There are 11 pair and a single adanal genital papillae. Quantitative measurements are given in Table 138.

Infective stage juveniles — The length of this stage is 759 (698 to 825) μm a = 28 (26 to 32); b = 9.6 (8 to 11); c = 10.1 (8 to 12.9).

Diagnosis — As Veremchuk[301] pointed out, *N. kirjanovae* shares similar characters with *N. glaseri* (e.g., lack of projection on male tail). However, the anteriorly placed excretory pore and the length of the infective stages clearly separates it from *N. glaseri*.

TABLE 139

Measurements of Females of *Neoaplectana belorussica*[301]

Character (N = 10)	Value	
	$\overline{X}$	Range
Total length (mm)	4.38	2.29—5.72
Length stoma	20	19—22
Length head to base of esophagus	218	194—234
Length head to excretory pore	75	
Length head to nerve ring	150	
% vulva	51	47—56
Length tail	42	32—51
Width at anus	53	
Length tail projection	4	3—6

TABLE 140

Measurements of Males of *Neoaplectana belorussica*[301]

Character (N = 10)	Value	
	$\overline{X}$	Range
Total length (mm)	1.53	1.18—1.91
Length stoma	14	11—14
Length head to base of esophagus	158	149—165
Length head to excretory pore	65	
Length head to nerve ring	120	
Length tail	27	24—30
Width at anus	35	
Length spicule	61	57—65
Length gubernaculum	44	41—51
Length tail projection	8	

Male specimens examined by the present author in Leningrad possessed a small tail projection, so this point still needs to be resolved. Veremchuk[301] did not include any further information on the biology or ecology of this nematode.

R. *Neoaplectana belorussica* Veremchuk 1969

This species was discovered near Minsk, in the Belorussian Soviet republic, from larvae of the elaterid *Athous niger* L.

1. Description[301]

Adults — In females, the cuticle is 2.7 μm thick. The vagina is 36 (32 to 41) μm long. The tail tip is drawn out to a small point. Quantitative values are given in Table 139. In males, the tail has a projection. There are 11 pair and 1 single genital papillae. Quantitative values are given in Table 140.

Infective stage juveniles — In this stage, the length is 525 to 675 μm, a = 24.3 to 25.6; b = 4.3 to 5.7; c = 10.8 to 11.5.

Diagnostic characters — Veremchuk[301] compares *N. belorussica* only with *N. bothynoderi* and no reference is made to the other described neoaplectanid species. From the lack of unique, specific characters, the present author regards this species as another population of *N. carpocapsae*. No information on the biology or ecology of this species is given with the original description.

S. *Neoaplectana hoptha* Turco 1970

This nematode was described by Turco[302] from material that had been maintained in the nematology collection at the U.S. Department of Agriculture, Beltsville, Maryland. It had originally been collected from the Japanese beetle *Popillia japonica* Newm.

It is unfortunate that Turco[302] did not clearly or adequately describe this species. Khan et al.[303] examined the paratype slides of *N. hoptha* male and female specimens from the U.S. Department of Agriculture nematode collection and noticed spicules and tail papillae similar to members of the genus *Heterorhabditis* and not to neoaplectanid nematodes. Khan et al.[303] assumed that the presence of long genital papillae in *N. hoptha* suggested that a bursa was also present. It is clear that *N. hoptha* is not related to any neoaplectanid nematode and belongs in the completely separate genus *Heterorhabditis*. See the latter genus for further discussion of this species.

T. *Neoaplectana semiothisae* Veremchuk and Litvinchuk 1971

This neoaplectanid was recovered from pupae of the geometrid *Semiothisa pumila* Kusn. in the Novosiberisk district of the U.S.S.R. Of 200 host pupae collected, 7 were infected with the above nematode.[304]

1. Description[304]

Adults — In females, the cuticle is finely annulated. The valve in the basal bulb is clearly visible. Eggs are spherical, 36×36 μm. The tail is short and pointed at the tip. Quantitative measurements are given in Table 141. In males, there are 11 pairs and a single adanal genital papilla. The spicular membrane does not extend to the tip of the spicule. The tail has a projection at the tip. Quantitative measurements are given in Table 142.

Diagnosis — The authors offer host selection as a distinguishing character separating this species from *N. feltiae* and *N. georgica*; however, this character is a doubtful one for neoaplectanid nematodes. The other character cited is how far the spicular membrane extends along the spicule — one third of the length of the spicule in *N. feltiae,* one ninth in *N. semiothisae* and to the tip in *N. georgica*. Just how variable this character is has not been determined.

VI. HETERORHABDITIDAE

This family contains the single genus *Heterorhabditis* Poinar[305] which is represented by several species or populations with a world-wide distribution (Table 143). Members of the genus *Heterorhabditis* are obligate parasites of insects which share with the neoaplectanids the specialized character of carrying around specific bacteria for nutrients. Characteristics of the family are as follows.

A. Heterorhabditidae Poinar 1976

Diagnosis: *Rhabditioidea* (Oerly) — The cuticle is smooth. There are six lips. Amphids are small and porelike. The stoma is vestigial. The pharynx is composed of an anterior cylindrical portion lacking a valve, an isthmus, and a terminal bulb containing a vestigial valve lacking bulb flaps. The male has caudal alae (bursa) supported by papillae. Ovaries are paired. They are pathogens of insects and are monogenetic. The generic diagnosis is as follows.

B. *Heterorhabditis* Poinar 1976

Diagnosis — There are six lips surrounding the mouth opening. Pro- and mesorhabdions are reduced and surrounded by pharyngeal tissue. Typical metastom and telor-

TABLE 141

Measurements of *Neoaplectana semiothisae* Females[304]

Character (N = 5)	Value	
	X̄	Range
Total length (mm)	3.24	2.05—5.65
Greatest width	340	
Length stoma	14	12—16
Width stoma	11	10—13
Length head to base of esophagus	182	156—221
Length head to excretory pore	65	
Length head to nerve ring	150	
% vulva	51	46—56
Length tail	22	13—28
Width at anus	58	

TABLE 142

Measurements of *Neoaplectana semiothisae* Males[304]

Character	Value	
	X̄	Range
Total length (mm)	0.47	0.43—0.50
Greatest width	35	
Length head to base of esophagus	112	
Length head to excretory pore	40	
Length head to nerve ring	72	
Length reflexion of testis	85	
Length tail	14	14—20
Width at anus	23	
Length spicule	42	36—44
Length gubernaculum	28	26—29
Length tail projection	2.4	

TABLE 143

Populations of *Heterohabditis* Collected from Nature

Species	Host	Host Family	Location	Ref.
bacteriophora Poinar	*Heliothis punctiger* Hall	Noctuidae	Australia	305
heliothidis (Khan et al.)	*H. zea* (Boddie)	Noctuidae	U.S.	303
hambletoni (Pereira)	*Eutinobothrus brasiliensis* (Hamb.)	Curculionidae	Brazil	306
hoptha (Turco)	*Popillia japonica* Newm.	Scarabaeidae	U.S.	302
Undescribed	*Sphenophorus coesifrons* (Gyll.)	Curculionidae	U.S.	307
No. 41088	*Pantomorus peregrinus* (Buch.)	Curculionidae	U.S.	308
Undescribed	*Ceutorrhynchus* sp.	Curculionidae	France	309
Undescribed	?	?	Australia	310
Undescribed	*Heteronychus arator* F.	Scarabaeidae	New Zealand	311
Undescribed	*Bothynus gibbosus* (DeGeer)	Scarabaeidae	U.S.	312

habdions are absent. The basal pharyngeal bulb lacks bulb flaps but contains an area with thickened linings enclosing a cavity or haustrulum. The cycle is heterogonic with both hermaphroditic females (arising from the infective stages) and dioecious females produced. They are amphidelphic with a median vulva. The genus is oviparous and/ or ovovivparous. The male has a single reflexed testis. Spicules are equal and paired. A gubernaculum is present. The bursa is open, normally with nine pairs of papillae.

The type species of the genus is *H. bacteriophora.* Its description follows.

C. *Heterorhabditis bacteriophora* Poinar 1976

1. Description

Diagnosis — The head is truncate or slightly rounded. There are six distinct protruding lips surrounding the mouth (in fixed specimens, which are often withdrawn or pointed toward the center of the mouth). There are also six labial papillae present. Amphids are located near the level of the labial papillae. Cheilorhabdions are represented as lightly refractile areas lining the anterior (noncollapsed) portion of the stoma. Metarhabdions have migrated anteriorly adjacent to the fused pro- and mesorhabdions, each metarhabdial segment bearing a small tooth. Telorhabdions are absent. A reduced collar is present. The basal pharyngeal bulb is often surrounded by the anterior portion of the intestine. The nerve ring surrounds the isthmus just anterior to the basal bulb. The excretory pore is usually posterior to the basal bulb. The intestine is composed of relatively few giant cells. Lateral fields and phasmids are inconspicuous. Females are variable in size. Ovaries are reflexed past the vulval opening. The hermaphroditic female has a vulva which is open and functional for oviposition (at least in the early stages). The female of the bisexual generation has a vulva which is nonfunctional for oviposition, often covered with a hardened deposit. The female has a pointed tail.

The male has a single reflexed testis leading into a seminal vesicle containing sperm cells. Vas deferens is well developed. Spicules are paired and separate. The shape of the capitulum is variable, from pointed to flat. The proximal portion of the gubernaculum is curved ventrally between the spicules. The bursa is peloderan, open, and supported normally by nine pairs of papillae — a small anterior pair, two pairs adjacent to the spicules, and six pairs distal to the anal opening. The latter six are in two sets of threes; rarely, the terminal group may consist of only two or even four papillae. More frequently, one or two of the three may be branched. Papillae pairs, four, seven, eight, and nine do not reach the rim of the bursa. Quantitative measurements of the adults are given in Tables 144 and 145.

Infective stage juveniles (third stage "dauer") (Figures 51 and 52) — The mouth and anus are closed. The pharynx and intestine are collapsed. The tail is pointed. The cuticle has longitudinal striae. Cells of a rodshaped bacterium occur in the ventricular portion of the intestine. The body is initially covered with the enclosing second stage cuticle, which is lost soon after the juveniles leave the host cadaver. The length is 570 (520 to 600) μm; greatest width 24 (21 to 31) μm; distance from head to nerve ring is 83 (81 to 88) μm; distance from head to excretory pore is 104 (94 to 109) μm; length of pharynx is 125 (119 to 130) μm; length of tail is 91 (83 to 99)μm.

Diagnosis — This nematode differs from other members of the Rhabditoidea by the following combinations of characters: six lips, vestigial stoma with reduced or modified rhabdions; terminal pharyngeal bulb lacking bulb flaps; males with a well developed papillate bursa; and "dauer" stages capable of entering hemocoel of healthy insects. The structure of the stoma and associated rhabdions appears to be unique in this species, and the heterogonic type of development makes this species unique among the insect parasitic rhabditids. The position of the excretory pore and length of the infective stage juveniles separate *H. bacteriophora* from other described species in the genus.

TABLE 144

Measurements of *Heterorhabditis bacteriophora* Females[305]

	Value			
	Hermaphroditic generation		Dioecious generation	
Character (N = 15)	X̄	Range	X̄	Range
Total length (mm)	4.03	3.63—4.39	3.50	3.18—3.85
Greatest width	165	160—180	190	160—220
Length stoma	8	6—9	7	6—9
Width stoma	8	6—9	7	6—9
Length head to base of esophagus	197	189—205	168	155—183
Length head to excretory pore	209	189—217	192	174—214
Length head to nerve ring	126	121—130	103	93—118
% vulva	44	41—47	47	42—53
Length tail	90	81—93	82	71—93
Width at anus	46	40—53	28	22—31

TABLE 145

Measurements of *Heterorhabditis bacteriophora* Males[305]

	Value — Dioecious generation	
Character (N = 15)	X̄	Range
Total length (mm)	0.82	0.78—0.96
Greatest width	43	38—46
Length stoma	3	2—4
Width stoma	2	1—3
Length head to base of esophagus	103	99—105
Length head to excretory pore	121	114—130
Length head to nerve ring	72	65—81
Length reflexion of testis	79	59—87
Length tail	28	22—36
Width at anus	23	22—25
Length spicule	40	36—44
Length gubernaculum	20	18—25

2. Bionomics

Life cycle[305] — *H. bacteriophora* was collected in Brecon, South Australia from diseased larvae of *Heliothis punctigera* Hall (Noctuidae). It was subsequently cultured on wax moth larvae *Galleria mellonella* and on artificial media in the laboratory.

Infective stage *Heterorhabditis bacteriophora* were capable of invading and killing larvae of *G. mellonella* in 48 hr. The nematodes were found in the midgut of the host from where they probably entered the body cavity. Entry through the spiracles and trachea is also a possibility. In 3 to 4 days after infection, the large hermaphroditic females produced young, which after two days had developed into males and females of the bisexual generation. After mating, the eggs hatched inside the females and developed to the second or even infective stage before emerging from the mother nematode.

Milstead and Poinar[313] reported that nematode development occurred within the temperature range of 16.8 to 29.5°C and that the LD_{20} for last instar larvae of *G. mellonella* was between three and six infective stages. A dose of 5 to 20 infective stages

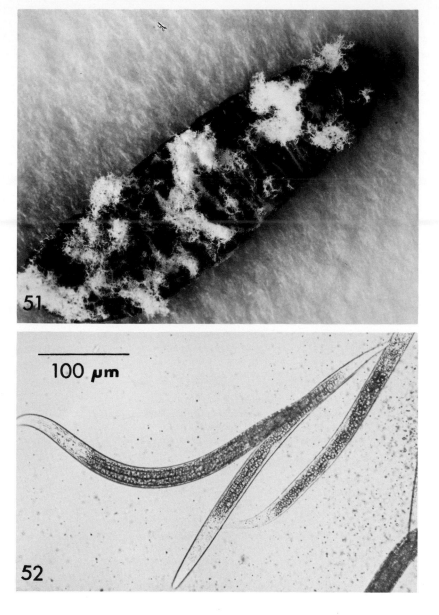

FIGURE 51 and 52. (51) Infective juveniles of *Heterorhabditis bacteriophora* clustered on the outside of their dead host. (52) Infective juveniles of *Heterorhabditis bacteriophora*.

resulted in the production of approximately 350,000 juveniles for each host cadaver. The juveniles have survived a 2.5% Ringer's solution for 14 months at 7°C.

Ecology — Some preliminary experiments were conducted by von Brecht[34] to determine the behavior of the infective stage juveniles of *H. bacteriophora*. To study the movement in soil, a series of six copper columns (each 5 cm deep and 8 cm in diameter) were filled with steam sterilized soil. Nematodes were placed either on the surface of the soil column or 6 in. below the surface with a pipette. After 1 to 2 weeks, the column was dismanteled, and the number of nematodes in each section was extracted with the Cobb sieving and gravity method and counted. Results of this experiment (Table 146) indicated that whereas the majority of the nematodes migrate upwards, some penetrate into the soil.

TABLE 146

Vertical Movement of *Heterorhabditis bacteriophora* Infective Stages in a 12-in. Column of Soil[314]

Height of soil (in.)	Nematodes applied to surface		Nematodes applied at a depth of 6 in.	
	Test A	Test B	Test A	Test B
10—12	328	780	622	37
8—10	70	13	77	3
6— 8	37	0	69	8
4— 6	46	0	63	3
2— 4	13	0	25	1
0— 2	0	0	112	4

TABLE 147

Hosts of *Heterorhabditis bacteriophora*[a,313]

Order	Family	Host species	Stages[b]	Infection[c]
Coleptera	Scolytidae	*Ips* sp.	L	E
Diptera	Culicidae	*Aedes sierrensis* (Ludlow)	L	E
		Culex tarsalis Coq.	L	E
Lepidoptera	Arctiidae	*Estigmene acraea* (Drury)	L	E
		Hyphantria cunea (Drury)	L	E
	Dioptidae	*Phryganidia californica* Packard	L	E
	Galleridae	*Galleria mellonella* (L.)	L	E
	Lasiocampidae	*Malacosoma californicum* (Packard)	L	E
		M. constrictum (Stretch)	L	E
	Liparidae	*Hemerocampa* sp.	L	E
	Noctuidae	*Heliothis punctigera* Wall.	L	N
		Pseudaletia unipuncta (Haworth)	L	E
		Spodoptera praefica (Grote)	L	E
	Notodontidae	*Schizura concinna* (J. E. Smith)	L	E
	Pieridae	*Colias philodice eurytheme* (Boisduval)	L	E
	Pyralidae	*Anagasta kuehniella* (Zeller)	L	E
		Paramyelois transitella (Walker)	L	E
	Tortricidae	*Archips argyrospila* (Walker)	L	E
Orthoptera	Blattellidae	*Blatella germanica* (Linn.)	N, A	E
		Supella supellectilium (Serville)	N, A	E

[a] Host supported *Heterorhabditis* development to invasive juvenile release.
[b] L, larva; N, nymph; A, adult.
[c] E, experimental infection; N, natural infection.

3. Host Range

Although the original host of *H. bacteriophora* was a noctuid moth larva (*Heliothis punctiger*), tests reported by Milstead and Poinar[313] show that representatives of four orders can successfully serve as hosts for *H. bacteriophora* (Table 147).

4. Culture

H. bacteriophora can be produced in larvae of *G. mellonella* or other suitable inects, using techniques similar to those described for *Neoaplectana*. Populations can also be grown on sterile dog food medium seeded with the symbiotic bacterium.

5. Bacterial Associates

The infective stage juveniles of *H. bacteriophora* contain cells of a specific bacterium in the lumen of their pharynx and intestine[315] (see Figures 53 and 54). These cells are liberated when the juveniles are placed in hanging drops of insect blood, and bacteria can be easily cultured on standard media. Examination of freshly attacked insects showed the bacterium present during nematode development. The bacteria turn the freshly killed insect larvae a reddish color, and the tissues take on a characteristic gummy consistency. The host is undoubtedly killed by multiplication of the bacteria associated with the nematode. The bacterial cells are large, peritrichously flagellated, motile, Gram-negative, nonsporeforming rods, occurring singly, in pairs, or rarely forming longer chains. Spherical cells are also formed, resulting from a disintegration of the cell wall. Results of cultural and biological studies are presented in the section on associated microorganisms.

6. Application

Although field trials have not been conducted with this nematode, laboratory tests showed that for controlling the red-humped caterpillar (*Schizura concinna*), approximately 2000 infective stage juveniles would have to be applied to a square foot of soil. This means that a single host cadaver could supply enough juveniles to treat 100 ft² of soil.[313]

D. *Heterorhabditis heliothidis* (Khan, Brooks and Hirschmann,) 1976

This species was collected from prepupae and pupae of *Heliothis zea* (Boddie) in North Carolina. It was originally described in the genus *Chromonema* but clearly belongs in the genus *Heterorhabditis*, which has priority over the former.

1. Description[303]

Adults — Lips are indistinct, each with one labial and one cephalic papilla. The stoma is short and wide. The pharynx has a simple corpus, indistinct isthmus, and basal bulb with a very weak valve. In females the lips are more prominent than in males, with cheilorhabdions represented as lightly scherotized areas lining the inside of the lip region. Pro-, meso-, meta- and telorhabdions are vestigial. A minute tooth-like structure is present in metastomal area. Ovaries are paired, opposite, and reflexed. First generation females are hermaphroditic. Three rectal glands are present. The anal lobe is present or absent. In males, the excretory pore opens slightly posterior to nerve ring. Spicules are paired, separate, and straight to slightly curved. Bursa is peloderan, open, and hyaline. There are three pairs of preanal and six pairs of postanal papillae. Quantitative measurements of the adults are presented in Tables 148 and 149.

Infective stage juveniles — Length is 644 (618 to 670) μm; greatest width is 25 (23 to 29) μm; length of the pharynx is 133 (130 to 139) μm; length of the tail is 108 (104 to 112) μm; the mouth is closed and the digestive tract nonfunctional.

Diagnosis — The position of the excretory pore and the length of the infective stage juveniles separate *H. heliothidis* from *H. bacteriophora*.

2. Bionomics[303]

Life cycle — Host death occurs 48 to 72 hr after exposure to the infective stage juveniles and is the result of a bacterial septicemia. The associated bacterium turns the

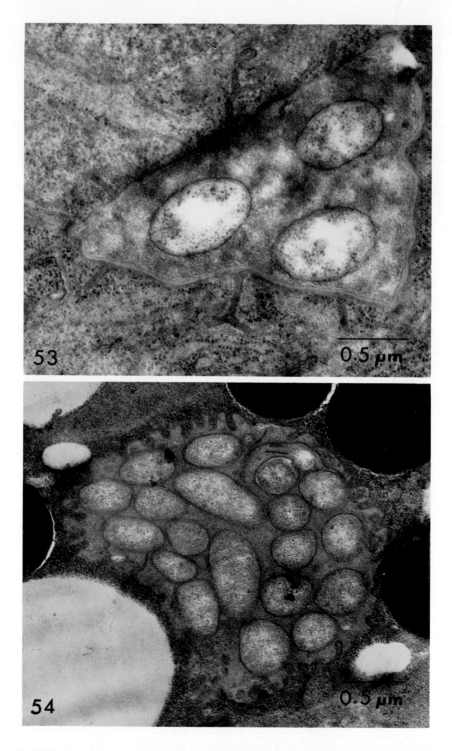

FIGURE 53 and 54.   (53) Cells of the symbiotic bioluminescent bacterium in the pharyngeal lumen of a *Heterorhabditis bacteriophora* infective juvenile. (54) Cells of the symbiotic bioluminescent bacterium in the intestinal lumen of a *Heterorhabditis bacteriophora* infective juvenile.

TABLE 148

Measurements of *Heterorhabditis heliothidis* Females[303]

Character (N = 25)	Value — First generation	
	X̄	Range
Total length (mm)	2.1	1.9—2.4
Greatest width	214	
Length stoma	6	
Width stoma	5	
Length head to base of esophagus	167	
Length head to excretory pore	286	
% vulva		46—53
Length tail	90	
Width at anus	33	

TABLE 149

Measurements of *Heterorhabditis heliothidis* Males[303]

Character (N = 20)	Value	
	X̄	Range
Total length (mm)	0.87	0.70—0.98
Length head to base of esophagus	137	
Length head to excretory pore	121	
Length head to nerve ring	105	
Length reflexion of testis	116	
Length tail	32—38	
Width at anus	23—32	
Length spicule	44	37—58
Length gubernaculum	24	21—25

host cadaver pink-red and finally a brick-red color. First generation hermaphroditic females are produced 3 days after host death. These hermaphrodites deposit eggs which produce second stage juveniles that in turn produce a bisexual generation of males and females. The young of these matings may develop into third generation females or into infective stage juveniles, depending on the amount of nourishment remaining in the host.

3. Host Range

Khan et al.[303] showed that *H. heliothidis* was capable of attacking a wide variety of insects, but that lepidopterous larvae were most susceptible. Susceptible hosts to this nematode are listed in Table 150.

4. Bacterial Associates

Khan et al.[303] reported that the infective stage juveniles of *H. heliothidis* were associated with a chromogenic, bioluminescent bacterium in a relationship similar to *N. carpocapsae* and *Achromobacter nematophilus*. Khan and Brooks[316] designated the bacterium from *H. heliothidis* as strain NC-19 and described it as a large asporogenous Gram-negative rod that grew well on nutrient agar. Further characteristics of this bacterium are presented in the section on associated microorganisms.

TABLE 150

Hosts of *Heterorhabditis heliothidis*

Order	Family	Host	Stage	Infection	Ref.
Coleoptera	Curculionidae	*Graphognathus* sp.	L	E	303
	Scarabaeidae	*Maladera castanea*	L	E	303
Diptera	Culicidae	*Culex pipiens* (L.)	L	E	303
Hymenoptera	Braconidae	*Apanteles militaris* (Walsh)	P	E	317
Lepidoptera	Arctiidae	*Estigmene acraea* Dru.	L	E	303
	Galleriidae	*Galleria mellonella* (L.)	L	E	303
	Noctuidae	*Helicoverpa virescens* Fab.	L	E	303
		Helicoverpa zea (Boddie)	P	N	303
	Sphingidae	*Manduca sexta* (Johann.)	L	E	303
	Tortricidae	*Argyrotaenia velutinana* (Walker)	L	E	303

TABLE 151

Measurements of *Heterorhabditis hoptha* Females[302]

Character	Value	
	X̄	Range
Total length (mm)	3.34	2.83—3.98
Greatest width	220	
Length head to base of esophagus	220	
Length head to excretory pore	180	150—225
Length head to nerve ring	180	
% vulva	47	43—49
Length tail	50	
Width at anus	62	

E. *Heterorhabditis hoptha* (Turco) 1970

This species was originally described by Turco[302] as a member of the genus *Neoaplectana*, but on the basis of an examination of paratypes of this species (Slides No. T-668p and T-1257p), the present author concludes that *hoptha* is really a heterorhabditid. This is in agreement with the conclusion of Khan et al.[303] after they examined the type material of this species. It was originally recovered from the Japanese beetle *Popillia japonica* Newm. in New Jersey in 1938.

1. Description[302]

Adults — In the female, the cuticle is smooth and the stoma reduced. The pharynx has a simple corpus, indistinct isthmus, and reduced terminal bulb. The excretory pore opening is at or near the nerve ring. The vulva is transverse with ventral protuberances (not shown in original description). The anus has a large postanal lip. The tail is conoid and pointed at the tip. In the male, the head has three lips and six labial and six cephalic papillae. The stoma is reduced. The excretory pore opening is at or near the nerve ring. The testis is single and reflexed. Spicules are paired and slightly curved. The gubernaculum is long and narrow. The tail is convex-concoid, with a rounded tip. There are 12 pairs of genital papillae — 4 pairs preanal, 2 pairs adanal, and 6 pairs postanal. Quantitative characters of the adults are presented in Tables 151 and 152.

Personal observations — Through the courtesy of Dr. M. Golden, the present author was able to examine some of the type material of *H. hoptha*. Slide No. T-688 which

TABLE 152

Measurements of *Heterorhabditis hoptha* Males[302]

	Value	
Character	X̄	Range
Total length (mm)	0.73	0.55—0.84
Length tail	27	
Width at anus	28	
Length spicule	47	43—60
Length gubernaculum	28	26—30

was labeled as containing both males and females contained only females. These females had the excretory pore opening at the level of the basal pharyngeal bulb or even at the level of the ventricular portion of the intestine (considerably lower than the level of the nerve ring). There were also six lips and not three as described by Turco, and the female tails ranged from pointed to rounded and only about half showed a postanal swelling.

The present author also examined paratype slide No. T-2175p of the males of *H. hoptha* and discovered that all three mature individuals possessed nine pairs of genital papillae that were never mentioned in the original description by Turco[302] (Figure 55). Although the bursal rays were present, the bursal membrane was absent. However, it was noted that most of the cuticle of these nematodes was etched (Figure 56). This etching of the cuticle was probably due to the corrosive action of certain solutions that were used fixing nematodes at that time. Substances such as sublimate or Flemming's solution, which contained corrosive elements, were used routinely as fixatives for nematodes at that time. Therefore, if these compounds could etch the nematode's cuticle, it is more than likely that they also destroyed the bursal membrane, which itself is cuticular in nature.

It is obvious that a careful redescription of this nematode is necessary since the present author has only given evidence that this species is a heterorhabditid, and serious omissions occurred in the original description. No biological data, other than the original host, were provided in the publications of this nematode.

F. *Heterorhabditis* sp. (Nematode 41088)

In 1941, a diseased larvae of *Pantomorus peregrinus* Buch. was collected at Saucier, Mississippi. It was discovered to be parasitized by nematodes which were accessioned under the number 41088. This nematode (referred to as Nematode 41088) was then collected from five other southeastern states from Louisiana to North Carolina. Although the nematode was never officially described and the only reference to it was presented in an unpublished work report from Gulfport, Mississippi,[308] it clearly falls into the genus *Heterorhabditis,* and the data merit recording in the present study.

1. Description[308]

Adults — The female has six lips which are usually retracted, each with a setaceous papilla. The stoma is reduced. The pharynx has a faint isthmus and terminal bulb. The excretory pore is located opposite to or slightly posterior to the pharyngeal bulb. A postanal lobe is present. The tail is short and stout, with an acute point. Ovaries are paired and opposed. The uterus has eggs and developing juveniles. The vulva is transverse. The male has a single testis. Spicules are sickleshaped, longer than the anal body width, and separate. The gubernaculum is half the length of the spicules. The

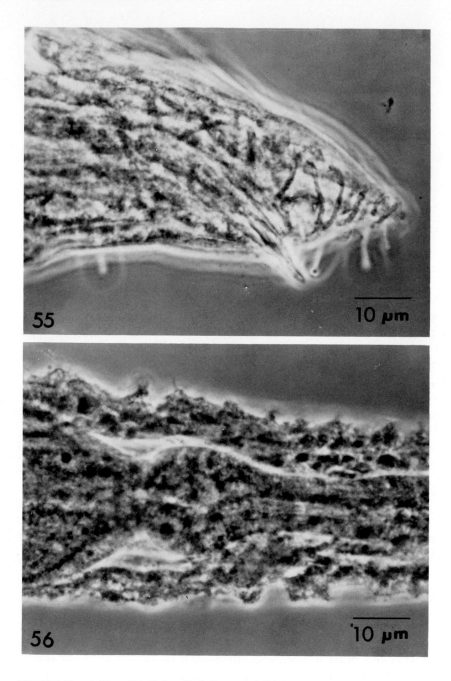

FIGURE 55 and 56.   (55) Male tail of *Heterorhabditis hoptha* (from paratype slide No. T-2175p of Turco) showing genital papillae. (56) Anterior region of an adult female *Heterorhabditis hoptha* (from paratype slide No. T-2175p of Turco) showing etched body wall.

bursa is peloderan and open, with seven or more pairs of genital papillae. The tail is short, stout, and acute at the tip. Quantitative measurements of the adults are presented in Tables 153 and 154.

Infective stage juveniles — This stage is about 0.5 mm long, enclosed in a second stage juvenile cuticle bearing longitudinal striae.

2. Bionomics[308]

Life cycle — The life cycle of Nematode 41088 was approximately 7 days at 70 to

TABLE 153

Measurements of the Female Nematode 41088 (*Heterorhabditis* sp.)[308]

Character (N = 5)	Value	
	X̄	Range
Total length (mm)	2.92	1.46—4.00
Greatest width	201	
Length stoma	9	
Length head to base of esophagus	260	
Length head to nerve ring	163	
% vulva	48	
Length tail	13	
Width at anus	53	

TABLE 154

Measurements of the Male Nematode 41088 (*Heterorhabditis* sp.)[308]

Character	Value	
	X̄	Range
Total length (mm)	0.92	0.74—0.99
Greatest width	57	
Length stoma	5	
Length head to base of esophagus	120	
Length head to nerve ring	76	
Length tail	39	
Width at anus	30	

75°F. The males were difficult to find, and the females were ovoviviparous. Two or more generations may occur in a single host, and the infective stages could be stored in soil or water for several months. The infective stages enter the host by way of the mouth or anus and occur in the body cavity 48 hr after initial exposure at which time the host dies.

Distribution — This nematode was recovered from 66 insect specimens from Louisiana, Mississippi, Alabama, Florida, Georgia, and North Carolina. Apparently, the nematodes were most active in the warmer months of the year.

3. Host Range

Nematode 41088 was collected from several insects in the field and proved infective to laboratory reared insects (Table 155).

4. Bacterial Associates

Littig and Swain[308] reasoned that the rapid death of the host accompanied by a sudden color change was due to bacteria associated with the nematode. They demonstrated that when body contents of insects attacked by Nematode 41088 were introduced into healthy insects, death resulted. There was no effect, however, when the contents were applied to the insect's cuticle or fed to the insect. Although the bacteria were never properly identified, the authors were undoubtedly dealing with the symbiotically associated bacterium of the *Heterorhabditis* nematode. In regards to the color change of a parasitized larvae, the authors noted that a pinkish tinge appeared

TABLE 155

Host range of Nematode 41088 (*Heterorhabditis* sp.)[308]

Order	Family	Host	Stage	Infec-tion[a]
Coleoptera	Curculioni-dae	*Listroderes obliquus* Klug.	L, P	E
		Pantomorus leucoloma (Boh)	L, P	N
		P. peregrinus Buch.	L, P	N
		P. striatus Buch.	L, P	N
	Scarabaeidae	*Pleurophorus* sp.	L	N
		Popillia japonica Newm.	L	E
Lepidoptera	Noctuidae?	Undetermined	P	E

[a] N, natural infection; E, experimental infections.

TABLE 156

Laboratory Control Studies with Nematode 41088 Against Beetle Larvae[308]

Host insect	No. insects used	Exposure period (days)	% Control
Pantomorus leucoloma	128	13—36	93 (68—100)
P. pergrinus	53	7—21	95 (89—100)
Listroderes obliquus	30	6	97 (95—100)

TABLE 157

Results of Adding Nematode 41088 to Soil in Flower Pots[308]

Host insect	No. insects used	Exposure period (days)	% Control
Pantomorus peregrinus	188	28	52
	125	73	27

in the host as early as a few hours after death. This color turned a pinkish burnt ochre the following day and eventually became brown, dark brown, and almost black.

5. Control Studies[308]

Laboratory studies — Larvae of white-fringed beetles were placed in 2-oz salve tins with the Nematode 41088. The results of these tests are shown in Table 156. Another test was conducted placing white-fringed beetle larvae in 10-in. clay pots outside. Soil containing "nematized" cadavers was added to the pots, and the results are shown in Table 157.

G. *Heterorhabditis hambletoni* (Pereira) 1937

This nematode was originally described in the genus *Rhabditis* by Pereira[306] and was found in Brazil parasitizing the cotton borer *Eutinobothrus brasiliensis* Hambleton. Little attention was paid to this early account until it was noted that this species fitted into the new genus *Heterorhabditis,* and the true pathogenic nature of these nematodes was proven. Ahmad discovered this species attacking white fringed beetles (*Graphognathus, Naupactus,* and *Pantomorus*) in Argentina.[381,382]

TABLE 158

Measurements of *Heterorhabditis hambletoni*
Females[306]

Character	Value
Total length (mm)	1.20—2.40
Greatest width	100
Length head to base of esophagus	110
Length head to excretory pore	90
Length head to nerve ring	80
% vulva	50
Length tail	90
Width at anus	28

TABLE 159

Measurements of *Heterorhabditis hambletoni* **Males**[306]

Character	Value
Total length (mm)	0.80
Greatest width	60
Length head to base of esophagus	130
Length head to excretory pore	100
Length head to nerve ring	90
Length tail	47
Width at anus	29
Length spicule	41
Length gubernaculum	20

1. Description[306]

Adults — The female cuticle is smooth. There are six lips. The pharynx has a cylindrical metacorpus and basal bulb. The vulva has protruding lips. Ovaries are paired and opposite. The tail narrows rapidly at the tip. They are oviparous or ovoviviparous. The male cuticle is smooth. The mouth has a minute vestibule. The pharynx has a posterior bulb lacking a valve. The testis is single and reflexed. The tail has a bursa and nine pairs of filiform papillae — three pairs of preanal papillae and six pairs of postanal papillae in two groups of three pairs each. Spicules are equal and joined at the tip. Quantitative measurements of the adults are presented in Tables 158 and 159.

Infective (third) stage juveniles — This stage has an ensheathing cuticle. The length is 467 μm (obtained from the scale in Figure 9 of Pereira's paper).

Diagnosis — Unfortunately, the single male described by Pereira[306] was in the process of molting, and a clear drawing of the male tail is lacking. Thus, clear diagnostic characters in *H. hambletoni* must wait until the species is redescribed, either from original material or a natural population. The only character that might be used to separate this species from the others in the genus is the relatively short size of the infective stage juveniles.

2. Bionomics[306]

Life cycle — Infective juveniles entered the host and produced only oviparous females (no males) which in turn laid eggs which produced juveniles that developed into males and females. The first generation females are oviparous, whereas the second generation females are ovoviviparous. Juveniles emerging from the second generation females develop into the infective stages.

Ecology — Nematodes were found only in dead cotton borers *Eutinobothrus brasiliensis*. Pereira[306] concluded that the nematode is of rare occurrence in nature since in a collection of 46 cotton borer larvae obtained in Marilia, only 1 dead larvae with nematodes was found. Attempts to culture the nematodes on decapitated cotton borer larvae placed on agar plates were largely unsuccessful. Also negative were attempts to cultivate *H. hambletoni* on decapitated *Musca domestica* larvae, blocks of egg albumin, and cotton roots. However, partial development did occur on decapitated cotton borer larvae, and Pereira obtained experimental infections using living cotton borer larvae.

VII. Tylenchida

Nematodes of the order Tylenchida all bear stylets in at least some stage of their development. They are well known as plant parasites, and several families are also associated with insects. Of the latter, the Neotylenchidae, Allantonematidae, and Sphaerulariidae contain representatives capable of destroying their hosts, and these nematodes can be considered as important biological control agents.

In contrast to the Steinernematidae and Heterorhabditidae, members of the above tylenchid families are mostly obligate parasites and cannot be cultured on artificial media. Insect-parasitic representatives of the Neotylenchidae do have a plant feeding generation, however, and it is possible to mass-produce the nematodes on the plant host.

Most species of allantonematids do not kill their host, and some seem to have little effect on the insect. Others, however, may reduce the host's life span or reproductive capacities, and a few are capable of causing insect mortality.

A. Neotylenchidae

The median bulb is reduced or absent, and a valve is lacking in members of this family. Those forms that parasitize insects possess two possible alternative life cycles, one on plants, the other in insects. Of the two insect parasitic genera in this family, namely *Fergusobia* and *Deladenus,* only species in the latter have been cultured and studied in regards to their use against insects.

One species of *Deladenus* attacks barine weevils which live in the stems of *Brassica napus* in France. After leaving the host, the juvenile nematodes can develop into two types of females. One is capable of reproducing in the body cavity of *Baris* larvae, while the second type multiplies and feeds in the plant stem. The nematode, which has not been described, reduces the life span and reproductive potential of the host.[318]

The other known insect parasitic species of *Deladenus* all infect *Sirex* woodwasps, their parasitoides, and occasionally other insects in the same habitat (Figures 57 to 59). Although several of the latter group have been described, most work, including field releases, has been done with *D. siricidicola* Bedding, and only this species will be covered in detail here. The above species was originally collected from eastern Europe but has a wide distribution.

1. Deladenus siricidicola[319] Bedding 1968
a. Description
Mycetophagous female (Figure 57) — The cuticle has fine transverse striae. Lateral fields vary from a fifth to a half of the body width, with 4 to 15 incisures. There are four lips, each containing a papilla. The stylet has well-developed basal knobs. The dorsal pharyngeal gland is greatly enlarged. Subventral glands are rudimentary, their ducts opening into a hollow chamber. The excretory pore is anterior to the hemizonid, usually at the level of the nerve ring. The prodelphic ovary is usually outstretched. There is no postvulval uterine sac. The tail is rounded.

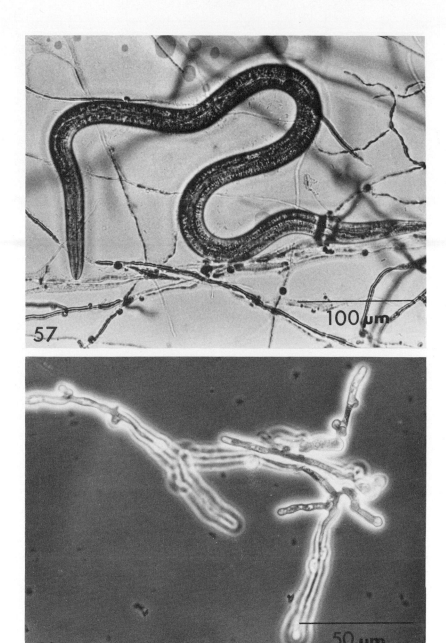

FIGURE 57 and 58.   (57) Mycetophagous female of *Deladenus* sp. (probably *siricidicola*)
feeding on the *Amylostereum* fungus. (58) Segments of *Amylostereum* removed from the
glands of a female *Sirex juvencus* from California.

Male — The testis outstretched. The bursa is peloderan, without rays. Spicules are
paired and separate. Spermatozoa may be large (10 to 12 μm) (found in mycetophagous
females) or small (1 to 2 μm) (found in insect-infective stage females).

Insect-infective stage female — This stage is much shorter and narrower than the
mycetophagous female. The cutice is striated. There are lateral fields over most of the
body. Lips are fused. The stylet is stout, lacking distinct knobs. Dorsal and subventral
glands are well developed. The excretory pore is distant from the hemizonid, the for-

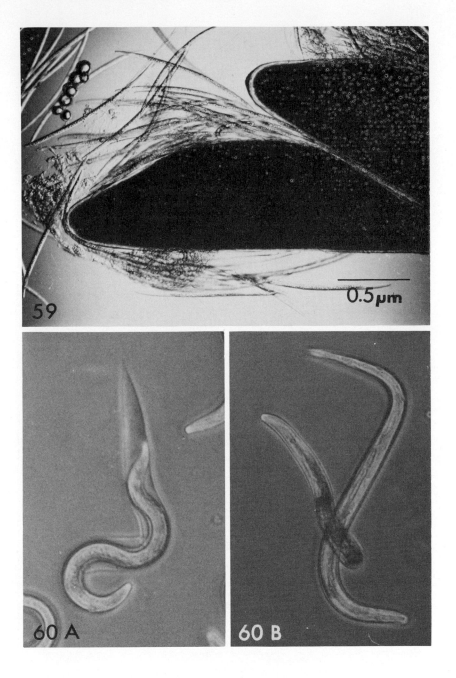

FIGURE 59 and 60.   (59) Eggs of *Sirex juvencus* containing juveniles of *Deladenus* sp. (probably *siricidicola*) from California. (60) *Howardula* sp. from *Carpophilus mutilatus* in California: (A) Molting juvenile; (B) Mating pair. (Courtesy of J. Lindegren.)

mer anterior to the nerve ring. The ovary is prodelphic, and a postvulval uterine sac is present. The tail is rounded.

Mature female parasites — This stage has an elongated (3 to 25 mm long, 100 to 200 μm wide) form. The tail is round or tapering. The cuticle is striated. The stylet is retracted into the head. The pharynx and glands are degenerate. The vulva is far posterior. The body contains developing eggs and hatched juveniles.

Qualitative measurements of the adult stages are given in Tables 160 and 161.

TABLE 160

Measurements of Mycetophagous and Insect-infective Stage Females of *Deladenus siricidi-cola*[319]

	Value			
	Mycetophagous		Insect-Infective	
Character (N = 50)	X̄	Range	X̄	Range
Total length (mm)	1.91	(1.50—2.71)	1.22	(0.80—1.60)
Greatest width	37	(39—45)	20	
Length stylet	10	(10—11)	21	(19—25)
Length head to base of esophagus	100		112	
Length head to excretory pore	98		83	
Length head to nerve ring	91		107	
% vulva	95	(93—96)	94	(93—95)
Length tail	43		35	
Width at anus	18		11	

TABLE 161

Measurements of Male *Deladenus siricidicola*[319]

	Value	
Character (N = 50)	X̄	Range
Total length (mm)	1.49	1.15—1.92
Greatest width	28	
Length stylet	10	10—11
Length head to base of esophagus	95	
Length head to excretory pore	91	
Length tail	47	
Width at anus	15	

b. Bionomics

Life cycle[320] — The life cycle of *D. siricidicola* and other species of this genus is quite extraordinary in that two separate self-perpetuating cycles are involved. In the mycetophagous life cycle, the nematodes are able to feed and develop on *Amylostereum aerolatum* (Fr.), the fungus symbiotically associated with the *Sirex* hosts of *D. siricidicola* (Figure 58). Juvenile nematodes removed from parasitized hosts matured to the adult stage within 5 days at 22°C on the fungus. After mating, each female may lay over 1000 eggs during its life span. Egg hatch occurs in 4 to 5 days at 22°C, and development from first stage juvenile to adult takes about 7 days. Fungal-feeding cultures of *D. siricidicola* have been continued for 4 years (around 100 generations) without an insect host.

As the substrate ages, some of the fungal-feeding juveniles develop into insect-infective stage females. These females are no longer capable of feeding on fungi and must find a susceptible insect host to continue their development. After finding a siricid larva, the infective females enter the hemocoel by direct penetration through the cuticle. The females may reach their full size within a few weeks after entering, but their reproduction system does not initiate development until the host begins pupation.

Eggs are produced about a week after host pupation and hatch in the reproductive system of the female nematode. Towards the end of pupation, the juveniles break out of the mother nematode, enter the host's hemocoel, and migrate to the host's reproductive organs.

TABLE 162

Developmental Hosts of *Deladenus siricidicola*

Order	Family	Host	Infection[a]	Ref.
Hymenoptera	Siricidae	*Sirex cyaneus* (Fab.)	N	320
		S. juvencus L.	N	320
		S. nitobei Mat.	N	320
		S. noctilio (Fab.)	N	320
		Urocerus sp.	E	322
		Xeris spectrum L.	N	320
Coleoptera	Melandryidae	*Serropalpus barbatus* (Schall)	N	320

[a]   N, natural infection; E, experimental infection.

Although the nematodes appear to have little effect on host larvae, they enter the developing eggs of the female hosts and suppress the developing ovaries. Thus, the female is sterilized with each egg containing 50 to 200 juvenile nematodes (Figure 59).

Nematodes in male *Sirex* spp. do not sterilize their hosts and usually die with the host, since there is no natural way for them to escape. In parasitized female hosts, the nematodes are passed out of the body during oviposition, either inside or along with the host eggs. They leave the host eggs, feed on the associated fungus, and develop into mycetophagous adults.

Natural occurrence — *D. siricidicola* has been recovered from 20 countries in Europe and Japan. It was introduced into New Zealand with the accidental appearance of the woodwasp *Sirex noctilio* prior to 1952.[321] Nematode cultures were also released in Australia in 1970 in a biological control program against *S. noctilio.* The major method of parasite distribution is through the dispersal activity of parasitized adult hosts.

c. Host Range

The natural host range of *D. siricidicola* includes siricids of the genera *Sirex* and *Xeris*, as well as a beetle of the melandryid family (Table 162). This nematode will also enter parasitoid larvae in the laboratory, but development in these hosts is never completed.[322]

d. Culture [323]

The most successful method of culturing *D. siricidicola* is on the symbiotic fungus *A. areolatum.* Cultures are established by dissecting the ooidial glands out of surface sterilized female hosts and streaking the contents on plates of potato-dextrose agar (PDA). After the fungus has started growing, juvenile nematodes are aseptically removed from insect hosts and placed on the agar cultures. Transfers can be made until a monoxenic culture is established. Nematode culture can be stored at 5°C in darkness for 6 to 12 months on PDA or corn meal agar. The nematodes and fungus will develop slowly under these conditions.

Because of expense, time, and handling difficulties, a flask method was devised for mass culture. The technique is to add 100g of wheat and 150 mℓ of water to a 500-mℓ Erlenmeyer flask which is autoclaved at 1.05 kg/cm² for 20 min. Before cooling, the flasks are shaken to separate the wheat and aerate the medium. After inoculation and incubation at 24°C in the dark for 4 to 6 weeks, water is added and the nematodes are removed, sieved, and allowed to settle. Using this method, 3 to 10 million nematodes can be recovered from each 500-mℓ flask. This amount is sufficient for inoculating about 100 m of timber.

There is no resistant stage in this nematode, and even in shallow water under an atmosphere of almost pure oxygen (5 to 10°C), the nematodes will only survive for several weeks. Thus, they are best maintained in agar cultures continuously and utilized when needed.

e. Field Application[323]

Since *D. siricidicola* is able to sterilize female *S. noctilio*, can be mass cultured in the free-living stage, and does not attack other insect parasitoids, it is well suited for use as a biological control agent. Releases have been made in Tasmania and Victoria. Introduction of *D. siricidicola* into wild host populations is done by inoculating *Sirex*-infested trees with nematodes, using a gelatin-based medium. A concentrated suspension of nematodes is mixed with a gelatin-water mixture (25,000 juveniles per 100 m*l* of gelatin) and introduced into special inoculation holes cut in the timber. The actual introduction is made with 50-m*l* plastic syringes, and for high levels of parasitism inoculations of 2000 nematodes were made at 30-cm intervals along one side of *Sirex*-infected logs or trees. The nematodes were also inoculated into small pieces of logs (billets) which in turn could be distributed in *Sirex*-infested timber.

The first experimental liberation was made in a 400-ha pine forest in northern Tasmania in 1970 and indicated that the nematode had great potential. Two years later, the nematodes had spread to 37% of the siricid-infested trees, and over 70% of the hosts emerging from nematode-infested trees were parasitized. In 1974, these figures were 70 and 90%, respectively.

One year after releasing nematodes in another forest in northern Tasmania, 90% parasitism was achieved, and over a period of 4 years the number of *Sirex*-killed trees dropped from 200 to 0.

D. siricidicola has also been liberated throughout most of the *S. noctilio*-infested areas in Victoria, and it is expected that a regular practice of release of nematodes and other parasitoids will be established.

Thus, *D. siricidicola* is a perfect example of a classical biological control program: selecting parasites of an introduced host from the land of origin and releasing them in a new location, where damage is most severe, in short, reestablishing a natural balance.

Of the seven described species of *Deladenus* that attack siricid woodwasps, detailed biological studies of only *D. siricidicola* have been reported, and only this species has been used for field applications. It is true that *D. wilsoni* Bedding can be cultured on two species of fungi symbiotically associated with *Sirex* species and is probably the most widely distributed species in the genus, but it has the serious drawback of attacking rhyssine parasitoids of siricids. In fact, most strains of *D. wilsoni* are found in rhyssine wasps and seem to prefer the latter over siricid hosts.

B. Allantonematidae

The allantonematids are stylet-bearing Tylenchida that are obligate insect parasites with no plant feeding generation like the neotylenchids. The fertilized free-living female enters the host by direct penetration through the cuticle and after developing into a mature parasitic female, deposits eggs or juveniles into the host's hemocoel. These juveniles mature to a stage that leaves the insect and completes its development in the environment. Most of the species in this family have been reported from Coleoptera and Diptera although fleas, thrips, and some bugs are also attacked.

There are two factors that make these nematodes difficult to use as natural control agents. First, many produce benign infections with little outright insect damage. Second, those species that do sterilize, reduce the life span, or— rarely — kill their host are sometimes difficult to maintain in the laboratory. For instance, some of the allantonematids attacking bark beetles are good pathogens, even killing their host. Yet maintaining the hosts in the laboratory is not practical, and in vitro methods of cultivation have not been established.

TABLE 163

Measurements of the Free-living Infective Females of *Heterotylenchus autumnalis*[324]

Character (N = 10)	Value
Length (mm)	0.98 (0.82—1.05)
Greatest width	25 (21—29)
Length stylet	21
Head to excretory pore	171
Head to nerve ring	144
Length tail	144
% vulva	76.8 (74.1—79.3)

In the following section, those species that have been used as biological control agents as well as those that have potential, i.e., adversely effect their host, and can be grown in the laboratory are discussed.

1. Heterotylenchus Bovien 1937

Members of the genus *Heterotylenchus* and *Psyllotylenchus* are unique in this family by having alternate gamogenetic and parthenogenetic parasitic generations. Whereas members of the latter genus parasitize fleas, a rather impractical host from the standpoint of laboratory maintenance, species of *Heterotylenchus* attack flies and beetles. One species has been cultured extensively in the laboratory and used as a biological control agent against *Musca autumnalis* in the U.S. It is discussed below.

2. Heterotylenchus autumnalis Nickle 1967

This species was first discovered in New York in 1966 as a parasite of the face fly *Musca autumnalis.* Its importance arises from its ability to sterilize female face flies.

a. Description[324]

Free-living infective female — The stylet is well developed. Knobs are lacking. The excretory pore is posterior to the nerve ring. The dorsal pharyngeal gland orifice is located about 21 μm behind the stylet base. Three prominent pharyngeal glands are present. The anus is faint and the tail is pointed.

Male — The pharyngeal glands are vestigial. The testis is single and outstretched. Spicules are paired and separate. The gubernaculum is absent. The bursa is small and faint. The tail is pointed.

Gamogenetic female — The body is coiled when killed. Lip region is slightly offset. The excretory duct, nerve ring, anus, and pharynx are degenerate. The intestine is well-developed. This stage is oviparous and the ovary is small and outstretched. Spermatheca is absent. The tail is pointed. Around 25 eggs are produced.

Parthenogenetic females — The body is straight when killed. The nerve ring and pharynx are degenerate. The anus is faint. The ovary extends almost to the head of the nematode, filling most of the body. This stage is oviparous. Hundreds of eggs are produced. The vulva is located near the anus. The tail is pointed.

Quantitative measurements of the above stages are given in Tables 163 through 165.

b. Bionomics

Life cycle[325] — After leaving the host, the free-living females and the males of *H. autumnalis* mate in the dung, and the fertilized female searches for a face fly larva. Entry into the host is accomplished by direct penetration through the cuticle. Afterwards, it develops into a gamogenetic female which deposits eggs directly into the

TABLE 164

Measurements of Male *Heterotylenchus autumnalis*[324]

Character (N = 10)	Value
Length (mm)	0.82 (0.75—0.90)
Greatest width	22 (19—30)
Length stylet	19 (18—19)
Head to excretory pore	160
Head to nerve ring	124
Length spicules	34 (30—36)
Length tail	118

TABLE 165

Measurements of the Gamogenetics and Parthenogenetic Females of *Heterotylenchus autumnalis*

Character (N = 10)	Gamogenetic females X	Range	Parthenogenetic females X	Range
Total length (mm)	5.18	(2.82—7.65)	1.44	(1.18—1.67)
Greatest width	168	(129—215)	120	100—148)
Length stylet	21		16	(15—17)
Length head to excretory pore	352		219	
Length head to nerve ring	200			
% vulva	82	(78—84)	92	(87—93)
Length tail	383		113	
Eggs	162 × 57		67 × 29	

hemocoel. These eggs hatch and produce juveniles which develop into smaller parthenogenetic females. They in turn deposit numerous eggs which hatch and develop into mature juveniles or even adult nematodes (especially males). These stages enter the reproductive system of the adult host, and in female face flies are passed to the outside through the oviduct during oviposition. The final stages of maturation and mating occur in the dung, thus completing the cycle.

Natural occurrence — First found in New York in 1964,[326] *H. autumnalis* had an incidence of infection of 23.1% in 14 counties surveyed. The nematode was subsequently found in Missouri, Montana, Nebraska, New Jersey, and Washington and probably occurs throughout the range of the host. Stoffolano found a *Heterotylenchus* sp. parasitizing the face fly in Denmark, and the present author found *H. autumnalis* in Swiss face flies; thus, the parasite probably was introduced into this country with its host in 1953.

Adult face flies collected from the field in Nebraska showed a 0 to 60% infection rate with *H. autumnalis*.[379] This parasite has also been reported in Czechoslovakia in several species of muscoid flies.[327]

Distribution of the nematode is insured by infected female hosts which, through "mock oviposition", deposit numerous free-living stages of *H. autumnalis* over the surface of the dung.

c. Host Range

Various dung breeding muscids have been found parasitized by *H. autumnalis* (Table 166), but *Musca autumnalis* is the only known host of this parasite in the U.S. Experimental infections of other Diptera, including *Musca domestica* and *Orthellia*

TABLE 166

Developmental Hosts of *Heterotylenchus autumnalis*
(All in the Family Muscidae)

Host	Infection[a]	Ref.
Hydrotaea meteorica L.	N	327
Morellia simplex Lw.	N	327
Musca autumnalis De G.	N	324
M. domestica L.	E	328
M. tempestiva Flln.	N	327
Orthellia caesarion (Mg.)	E	328

[a]   N, natural infection; E, experimental infection.

caesarion, were conducted by Stoffolano and Streams,[328] but host reactions usually prevented the nematodes from completing their development. Occasionally, the nematode was able to complete its development in *M. domestic* and *O. caesarion* but was unable to leave the host.

d. Culture

Stoffolano[329] devised several methods for culturing *H. autumnalis* in the face fly host, and the most efficient method is described here. Essentially, it involves artifically seeding the manure to obtain infection of fly larvae. Infected adult flies are anesthesized and placed in a blender for 15 sec. The contents is then poured in a Baerman funnel lined with tissue paper. The juvenile nematodes crawl through the paper and are collected in the tube at the base of the funnel. They are then placed on fresh manure and stored for 24 hr, after which mating has occurred and the fertilized females are ready to enter hosts. Healthy fly larvae are placed in the manure for 3 to 4 hr, then removed and maintained on fresh manure.

e. Field Trials

Since *H. autumnalis* normally sterilizes female face flies and is naturally spread by the infected hosts to new environments, it seemed to be a natural biological control agent. To date, it has been released at least once.[330,331] Some 10,000 nematode-infected pupae of *M. autumnalis* were released in northern California from the coast in Humboldt County to the mountains in Modoc County to stem the tide of this recently introduced pest. Face flies collected in the Bay Area by the present author were found to be infected by *H. autumnalis*.

Another *Heterotylenchus,* not yet determined to species, attacks the Australia bush fly *Musca vetustissima.* The morphology and life cycle of this nematode, which has been cultured on laboratory-reared bush flies, is similar to *H. autumnalis.* The parasite occurs in all five eastern Australia states as well as Tasmania, and the incidence of parasitism reached 30%. In experimental infections, the parasites prevented egg development, although the nematode has never been released in attempt to control the bush fly in the field.[332]

3. Howardula Cobb 1921

Representatives of the allantonematid genus *Howardula* also cause sterility of their insect hosts. Characterized by the absence of a stylet in the free-living male stage, members of this genus parasitize Coleoptera, Thysanoptera, Diptera, and even Acarina.

4. Howardula husseyi Richardson, Hesling, and Riding 1977

This nematode parasitizes the mushroom phorid *Megasalia halterata* in England and probably other countries.

a. Description[378]

Free-living infective females (fourth stage) — These slender nematodes have very fine transverse striations. The lateral field is 6 to 7 μm wide. Lips are fused, and there are three pharyngeal glands. An excretory pore is absent. The anus is rarely visible. The ovary is outstretched. The vulva is posterior. The tail has a mucron. The length is 426 to 529 μm. The width is 18 to 22 μm. The length of the stylet is 13 to 19 μm. The position of the vulva from the anterior end of the nematode is 79 to 86%.

Free-living males — This stage has very fine transverse cuticular striations. The lateral field is 2 to 3 μm wide. A stylet is absent. The pharynx is vestigal. Spicules are paired and equal. A gubernaculum is absent. The bursa is peloderan. The testis is extended. The tip of the tail is rounded. The length is 356 to 457 μm; the width is 16 to 20 μm; the length of the tail is 26 to 41 μm; the length of the spicules 10 to 13μm.

Mature parasitic females — The body is swollen. This stage is polymorphic. Stylet, pharynx, and anus are absent. The vulva has lips. They are oviparous (rarely ovoviviparous) and occur in pupa and adult flies. The length is 529 to 1109 μm; the width is 65 to 211 μm. The position of the vulva from the nematode is 86 to 96%.

b. Bionomics[333,378]

Life cycle — The fertilized free-living females of *H. husseyi* enter the hemocoel of *M. halterata* larvae by direct penetration through the cuticle. They molt within 24 hr after reaching the insect hemocoel and begin developing into mature parasitic females. The parasites attain maximum size during host pupation and begin depositing eggs when the pupae are 5 to 6 days old. The hatching juveniles molt once in the fly, and the second stages penetrate the host's ovaries, enter the oviducts, and are liberated during fly oviposition. Those nematodes in male hosts never mature completely and die soon after their host perishes.

After reaching the substrate, second stage juveniles may molt twice to fourth stage females or three times to adult males; however, the initial cuticles are retained as sheaths around the nematodes until the last molt.

Natural occurrence — The parasite was recovered from several mushroom farms in Surrey, Berkshire, Hampshire, and West Sussex in England. Depending on the time of year and number of flies present, parasitism varied between 0 and 75%. The nematode is distributed by parasitized female flies.

c. Host Range[333]

Thus far, *Megaselia halterata* Wood (Phoridae: Diptera) is the only known natural host of *H. husseyi,* and experimental infections of other insects have not been reported. Aside from reducing or destroying the fat body of the host, the parasite also limits its ability to copulate and lowers egg production. Ten or more parasites in a single fly prevented reproduction.

d. Application

The host *M. halterata* can be maintained in the laboratory on spawned compost and has a life cycle of 21 days. The host larvae feed on fungi and can destroy the mycelial growth of mushrooms. Flies brought into the laboratory from commercial mushroom farms with 3.7 to 8.5% parasitism had more than 90% parasitism after three generations in breeding cages.[333] This shows that parasitized flies can be maintained in the laboratory for eventual release. Hussey[334] felt that if the parasites could be uniformly distributed artificially through the compost before the host eggs hatched, complete parasitism could be achieved. This would roughly be equivalent to the natural dispersal of two or more parasitized female hosts per 30 g of compost.

Although there is no truly resistant stage in the life cycle of this nematode, and therefore storage of the parasite would present a problem, dispersal could be done with parasitized female hosts that would release the nematodes naturally. By increasing the natural populations of *H. husseyi* through the release of infected adults, a higher incidence of hyperparasitized host larvae could be attained, resulting in a higher number of sterilized adults in succeeding generations.

5. Howardula medecassa Remillet and van Waerebeke 1975

Remillet and van Waerebeke[335] described two species of *Howardula* from *Carphophilus* beetles in Madagascar and oceanic islands. The parasites were responsible for atrophied gonads in the adult beetles. An almost identical species of *Howardula* was discovered by Lindegren attacking *Carpophilus mutilatus* in California (Figures 60 and 61).

The biology of all three species is very similar and will be discussed below, although only the description of *H. madecassa* Remillet and van Waerebeke[335] will be given since all three nematodes resemble each other so closely. The Californian species or strain has not yet been described, whereas the second species from Madagascar, *H. truncati* Remillet and van Waerebeke,[335] occurs in a different host from *H. medecassa.*

a. Description

Infective stage females — The body is relatively thick and straight at death. The cuticle has fine transverse striae. The head is flat at the tip and is not separated from the body. There are four faint lips. Papillae are not visible. The tail is terminated by a mucron 2 μm long. The stylet is thick. There are three pharyngeal glands. The anus is faint.

Mature parasitic female — This stage is bent dorsally when heat-killed. The cuticle has fine annulations. Head and tail are rounded and lips are indistinct. A stylet is present. Pharyngeal glands are atrophied. The intestine and anus are degenerate. The ovary is reflexed. The vulva is distinct. This stage is ovoviviparous.

Males — This stage is bent ventrally when heat-killed. The cuticle has fine transverse striations. The head is blunt. The tail is rounded. Lips, papillae, and stoma are not visible. Stylet and pharyngeal glands are degenerate. The testis is outstretched. Spicules are paired and equal. The gubernaculum is small. The bursa is absent.

Quantitative values of the adults are presented in Tables 167 and 168.

b. Bionomics[335,336]

Life cycle (applies to *Howardula* species of *Carpophilus* beetles) — Beetles of the genus *Carpophilus* breed in fallen fruit continuously throughout the year, depending on the temperature and availability of food. The infective stage female of *Howardula* penetrates directly through the cuticle of the host larva. From one to six infective stages may enter a single host in nature (depending on the nematode species). When the nematode enters a first stage host larva, its growth and maturation are completed when the host pupates. Then the female begins to deposit juveniles in the body cavity of the host and continues doing so throughout the life of the host. A single female of *H. madecassa* can deposit about 3400 juveniles into the host's hemocoel. The first molt occurs in the egg, and final development to the third stage juveniles continues into the adult stage of the host. The mature third stage juveniles penetrate the intestine, enter the digestive tract, and are passed out of the host's rectum into the environment.

In *H. madecassa,* the juvenile nematodes begin to leave the host 8 to 16 days after

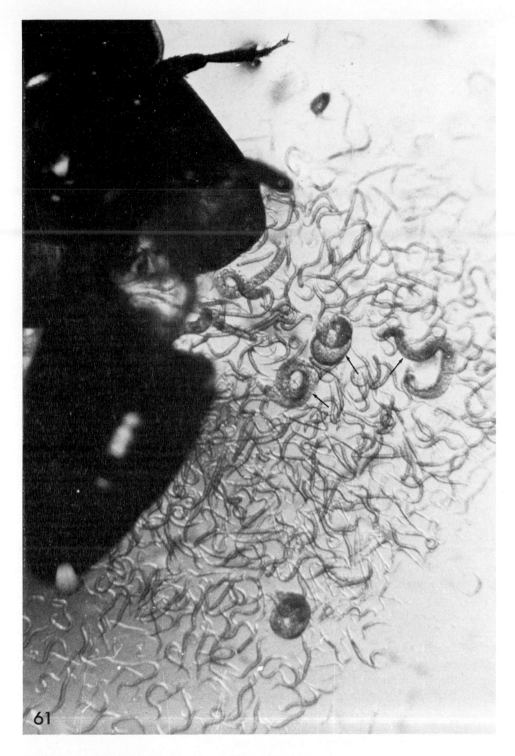

FIGURE 61.   A parasitized adult *Carpophilus mutilatus* beetle opened to show the adults (arrows) and juveniles of *Howardula* sp. (Courtesy of J. Lindegren.)

TABLE 167

Measurements of the Infective Stage and Mature Parasitic Females, Respectively, of *Howardula madecassa*[335]

Character	Value	
	Infective stage	Mature parasite
Length (mm)	0.387 (0.370—0.396)	1.83 (1.52—2.74)
Greatest width	19—21	150—200
Head to excretory pore	41 (36—43)	—
Length of stylet	22 (21—22)	—
Head to nerve ring	78 (73—82)	—
Length tail	22—26	26—32
% vulva	88 (87—88)	97 (95—98)

TABLE 168

Measurements of Male *Howardula madecassa*[335]

Character (N = 10)	Value
Length (mm)	0.416 (0.362—0.460)
Greatest width	15—21
Head to excretory pore	43 (39—47)
Head to nerve ring	59 (50—70)
Length spicules	12—13
Width spicules	5
Length gubernaculum	4
Length tail	31—39

the beetle has reached the adult stage. A double molt to the adult stage occurs in the nematode juveniles between 24 and 48 hr after they have left their host. The males die after mating and the fertilized females can survive up to 2 weeks in water.

The mature females of *H. truncati* become surrounded with host tissues in the adult beetle and eventually burst, releasing the juveniles into the hemocoel.

Natural occurrence and host range — *H. madecassa* was found parasitizing *Carpophilus fumatus* Boh. in Madagascar and on the island of Mayotte in the Indian Ocean. It also parasitizes *C. notatus* Murray in Madagascar. Rates of parasitism in the former host in Madagascar varied between 10 and 43% depending on the season and location. *H. truncati* parasitizes *C. truncatus* Murray on the atolls, Farguhar and Providence in the Indian Ocean. The rate of parasitism reached 18% in the former location. The *Howardula* sp. from California was found only in *C. mutilatus* Erich, but parasitism occurred in Alabama, Mississippi, Georgia, Texas, Florida, and Mexico. The nematodes are easily distributed by parasitized adult hosts to new areas.

Parasitism of *C. truncatus* by *H. truncati* resulted in atrophy of the gonads when more than two adult female nematodes were present. *C. mutilatus* females parasitized by *Howardula* sp. produced only 20% of the number of juveniles and 67% of the number of mature eggs of nonparasitized beetles.

c. Culture and Potential Application

Lindegren[336] maintained parasitized colonies of *C. mutilatus* in quart jars one third full of 1:1 mixture of sand and peat moss. Autoclaved moist figs were added weekly as a food source, along with water to keep the substrate moist. Colonies were maintained for up to 1 year without the addition of healthy beetles.

The effect of *Howardula* spp. on *Carpophilus* beetles is to lower the insect's repro-

TABLE 169

Measurements of the Infective Stage and Mature Parasitic Females of *Howardula benigna*[337]

	Value	
Character	Infective stage	Mature parasite
Length (mm)	0.57	3.5—5.0
Greatest width	26	164
Length stylet	25	—
Head to nerve ring	86	18
Length tail	34	35
% vulva	90	98

ductive ability. Only rarely is there host mortality, and the nematodes seem to have a restricted host range. However, Lindegren[336] remarked that the introduction of *Howardula* sp. into areas such as fruit dumps might reduce the number of beetles and restrict their ability to disseminate.

The type species of the genus *Howardula, H. benigna,* attacks beetles of the genus *Diabrotica* in the U.S. and probably elsewhere.[337] Since *Diabrotica* beetles are serious crop pests and the nematode affects the reproductive potential of the the host, it is listed here as a possible biological control agent.

6. Howardula benigna Cobb 1921

a. Description[337]
Infective stage female — The cuticle has transverse striae. Lips are fused. Labial papillae are not visible. Three pharyngeal glands and a stylet are present. The head is flat. The tail is conical. The excretory pore is anterior to the nerve ring.

Mature parasitic female — The body cavity is nearly filled with the uterus containing developing eggs and juveniles. The anus is vestigial. The vulva is sometimes terminal. This stage is ovoviviparous. The ovary is reflexed.

Male — The cuticle has faint striae. Lips are fused. Stylet and pharynx are vestigial. Spicules are paired and nearly straight with distal ends blunt. The gubernaculum is simple and narrow. Lateral cuticular wings are present. The excretory pore is anterior to the nerve ring. The tail is conical.

Quantitative measurements are given in Tables 169 and 170.

b. Bionomics[337]
Life cycle — Up to 13,000 juveniles of *H. benigna* have been removed from the body cavity of a single *Diabrotica* adult. The mature juveniles enter the oviducts of females beetles and are deposited on the soil with the eggs. Once in the soil, the nematodes molt and mate, and the fertilized females search out *Diabrotica* larva. After entering the host's hemocoel (probably by direct cuticular penetration), the nematodes swell up into an egg-producing sac and deposit juvenile nematodes into the host's body cavity. As occurs in other members of this species, the eggs hatch in the uterus of the mother nematode.

Natural occurrence — *H. benigna* was first discovered parasitizing *Diabrotica* beetles in Ohio. Subsequent surveys involving about 1500 specimens of *D. vittata* showed the parasite to be common throughout the northeastern, southern, and southwestern states, including Michigan, Massachusetts, Virginia, Iowa, Illinois, Maryland, Mississippi, Minnisota, Louisiana, North Carolina, Connecticut, Alabama, and California.

TABLE 170

Measurements of Male *Howardula benigna*[337]

Character	Value
Length (mm)	0.61
Greatest width	25
Length stylet	15
Head to nerve ring	79
Length spicules	18
Length gubernaculum	6
Length tail	41
Width at anus	19

It was also recorded in Nova Scotia and Ontario, Canada. The rate of parasitism ranged from 8 to 70% in the above states. The nematodes are distributed by ovipositing adult female beetles.

c. Host Range[337]

H. benigna parasitizes *Acalymma* (syn. *Diabrotica*) *trivattatum* (Mann), *A. vittatum* (Fab.), and *Diabrotica unidecempunctata howardi* Barber in the U.S. In Europe, *H. phyllotretae* Oldham occupies the same niche as *H. benigna* and parasitizes flea beetles of the genus *Phyllotreta*. Beetles parasitized by *H. benigna* are smaller and lighter than nonparasitized individuals. Heavy infections may kill the host larvae and partially or completely sterilize the adult beetle.

d. Application

Although there have been no reports of distributing *H. benigna* in the field for controlling cucumber beetles, it is not difficult to rear the host under laboratory conditions, and field releases could be made. Although the free-living infective stage of *H. benigna* may only survive in soil for several weeks, the nematodes could be introduced into new areas through the release of infected adult beetles.

VIII. SPHAERULARIIDAE

Members of this small family, comprising three genera, are characterized by the possession of only two pharyngeal glands. Representatives of two genera contain parasitic females that evert their reproductive systems into the body cavity of the host. All representatives are restricted to one family of hosts. All the species in the Sphaerulariidae are capable of either killing the larval stage of the host or sterilizing the adult stage.

Sphaerularia bombi Dufour is the only species in the genus and occurs throughout various parts of the world as a parasite of queen bumblebees.[338] The cycle is difficult to maintain in the laboratory, and the exact method of nematode entry into the host is still a mystery.

Scatonema wulkeri Bovien[339] is a parasite of scatopsid dung flies and has been reported only once in Denmark. Continuous laboratory cultures of this nematode were never maintained but should not be difficult if it ever was necessary to control scatopsid flies.

The most studied species is *Tripius sciarae*, which parasitizes members of the Sciaridae. Since these gnats are often a nuisance and control possibilities with this nematode are quite promising, *T. sciarae* is discussed further here.

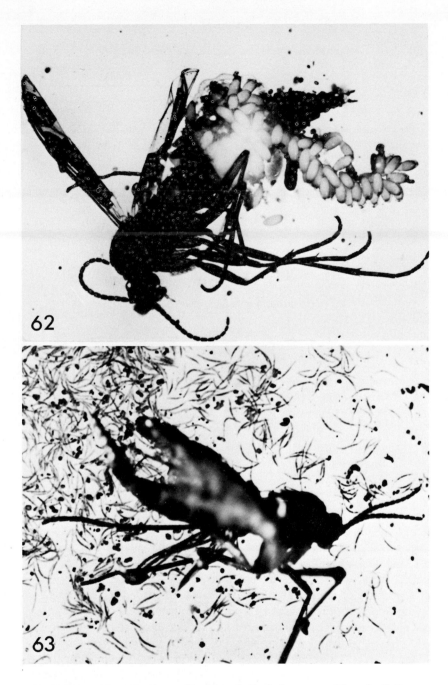

FIGURE 62 and 63. (62) Unparasitized female *Bradysia paupera.* (Photo by C. Doncaster.) (63) Female of *Bradysia paupera* sterilized by *Tripius sciariae.* (Photo by C. Doncaster.)

A. *Tripius sciarae* Bovien[340] (syn. *Proatractonema sciarae* Bovien 1944)

This nematode was first discovered in Denmark but is probably worldwide in distribution. It parasitizes flies of the family Sciaridae (Mycetophilidae) and can be easily reared under laboratory conditions. Because most parasitized host larvae die before reaching the pupal stage and all adult flies are sterilized by the nematode, it is a potential biological control agent (Figures 62 and 63).

TABLE 171

Measurements of Infective Stage Female *Tripius sciarae*[340,341]

Character (N = 25)	Value	
	$\overline{X}$	Range
Total length (mm)	0.378	(0.350—0.400)
Greatest width	15	(14—20)
Length stylet	10	(9—14)
Length head to bifurcation of pharynx	26	(24—29)
Length head to excretory pore	65	(62—72)
Length head to nerve ring	26	(24—29)
% vulva	66	(65—72)
Length tail	70	(59—80)
Eggs	54—59	× 22—59

TABLE 172

Measurements of Male *Tripius sciarae*[340,341]

Character (N = 25)	Value	
	$\overline{X}$	Range
Total length (mm)	0.379	(0.350—0.394)
Greatest width	16	(14—20)
Length head to excretory pore	69	(54—77)
Length head to nerve ring	54	(47—60)
Length tail	61	(51—70)
Width at anus	13	(11—16)
Length spicules	12	(10—13)
Length gubernaculum	4.3	(2.9—5.7)

1. Description[340,341]

The genus *Tripius* is characterized by the partial or complete extrusion of the uterus through the vulva of the parasitic female; however, the everted uterus is never longer than the female nematode. Also, the eggs never hatch inside the body of the female nematode. There are two species in the genus, *T. sciarae* and *T. gibbosus;* the latter species is a parasite of cecidiomyid larvae and was recorded once from Germany.[342]

Free-living females — This stage still retains the last juvenile cuticle which plays an important role in the process of penetration. Stylet and penetration glands are well developed. The intestine and anus are vestigial and probably nonfunctional.

Free-living male — This stage is also enclosed in the last juvenile cuticle. The stylet, intestine, and pharyngeal glands are vestigial. Spicules are paired and separate.

Quantitative measurements of the adults are given in Tables 171 and 172.

Mature parasitic females — The body is swollen and is approximately 900 μm long and 280 μm wide. Uterine cells are partially or completely expelled through the vulva into the host's body cavity. The intestine is degenerate. The body cavity is filled with developing eggs.

2. Bionomics[340,341,343]

Life cycle — The infective stage female of *T. sciarae* enters the host by direct penetration through the cuticle (Figure 64). It first attaches itself to the host by dissolving the anterior portion of its retained last stage juvenile cuticle with pharyngeal gland

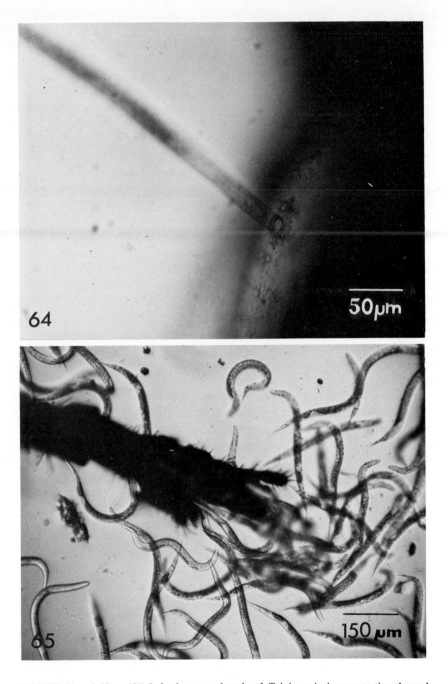

FIGURE 64 and 65.   (64) Infective stage female of *Tripius sciariae* penetrating through the cuticle of a sciarid larva. (Photo by C. Doncaster.) (65) Parasitic juveniles of *Tripius sciariae* being expelled through the ovipositor of an infected *Bradysia paupera*. (Photo by C. Doncaster.)

secretions. After attachment, further gland secretions and the stylet are used to puncture a small hole in the host's cuticle. The nematode then enters the hole and leaves the juvenile cuticle attached to the host's body. Only the females enter the host and only after they have been fertilized. Development continues in the host's homocoel, and after 7 days the now swollen females begin depositing eggs into the host's body cavity. These eggs hatch in 3 days at room temperature, and the juveniles develop to the fourth stage. They may leave the host larva through a small rupture in the latter's intestine. The fly usually dies within the next few days, and the remaining nematodes make their exit.

After leaving the host, most fourth stage juveniles matured to the adult stage in 2 days but retained the juvenile cuticle. The males died after mating and lived only 7 to 10 days after emergence from the host. The infective stage females then searched for a host but remained viable for only 2 weeks in in moist soil. *T. sciarae* could also be carried into the pupal and adult stage of the host.

Natural occurrence — Bovien[340] found 30% of the *Sciara* larvae associated with *Convallaria majis* roots to be infected with *T. sciarae* in Denmark.

Wachek[344] reported the same nematode from sciarids in Germany, and Poinar[341] described their occurrence in British sciarids. The same species has been collected from California sciarids by the present author. *T. sciarae* can probably be found in most areas containing scarid flies and are efficiently dispersed by parasitized adult hosts, which continue their ovipositional behavior and deposit small packets of nematodes over the soil surface (Figure 65).

T. sciarae has only been recovered from two species of the dipterous family Sciaridae, *Bradysia paupera* Tuom. and *B. impatiens* (Johann.). Bovien[340] simply listed *Sciara* sp. as the host of this parasite in Denmark.

3. Culture

Populations of sciarids can be easily maintained in compost or on fungi growing on agar plates. Infection with *T. sciarae* readily occurs when fly larvae are placed on soil containing infective stage nematodes. The hosts can then be removed to culture chambers until the nematodes are ready for emergence. The infective stage females of *T. sciarae* are probably the most resistant stage of the parasite and, unfortunately, only survive about 2 weeks at room temperature. However, their life may be prolonged at lower temperatures, and the lack of a durable resistant stage is partially compensated for by the relative ease of maintaining the infection in sciarid hosts.

4. Application

Soil containing infective stage females of *T. sciarae* was distributed over several flats in one of the Rothamsted greenhouses.[341] After 4 weeks, the sciarid populations in that greenhouse had dropped drastically and the few adult flies recovered were all parasitized. These results show the effectiveness of the parasite and how quickly they are distributed (by parasitized adult hosts) throughout a restricted area (Figure 65). They would make ideal biological control agents in confined areas, such as mushroom houses and greenhouses.

Chapter 5

MICROORGANISMS ASSOCIATED WITH ENTOMOGENOUS NEMATODES

Those microorganisms which are considered mutualists or inquilinists are discussed here; those which are pathogenic are discussed under the section covering enemies of entomogenous nematodes.

I. MUTUALISTIC ASSOCIATIONS

Thus far, only bacteria and fungi are known to form mutualistic associations with nematodes. Relationships with the former group of microorganisms are known only for members of the nematode families Steinernematidae and Heterorhabditidae (Table 173).

The bacterium isolated from strains of *Neoaplectana carpocapsae* was characterized and described as *Achromobacter nematophilus*.[245,246] Undoubtedly, its taxonomic position will be modified in the future based on recent views on bacterial nomenclature; however, it will be referred to here as the above species. The relationship between *A. nematophilus, N. carpocapsae,* and the insect host is discussed in the section on *Neoaplectana.* The bacterium is carried from insect to insect in the intestine of the infective stage juvenile nematodes.

Cultural and biological characteristics of two strains of *A. nematophilus* isolated from American and Soviet populations of *N. carpocapsae* are presented in Tables 174 and 175. The bacterium is a medium to long-shaped rod suggestive of the vegetative cells of *Bacillus* (Figure 45). Although it is an asporulating species, cultures of *A. nematophilus* form characteristic spheres, along with the normal rods. Freeze-etching studies showed that these spheres form when the outer bacterial wall is dissolved leaving only the cell membrane intact (Figure 66). Other investigations, related to the chemical characteristics of this bacterium, are mentioned below.

In a serological study comparing *A. nematophilus* isolated from various strains of *N. carpocapsae* with similar bacteria isolated from other *Neoaplectana* species and three standard bacterial species, Parvez[345] made some interesting observations. Using antiserum prepared against the purified flagellar protein from the DD-136 strain, Parvez showed that the bacteria from the Agriotos and Mexican strains shared some common antigenic sites with the bacteria from the DD-136 strain. He regarded the bacteria originating from these three strains as a single group of apparent specificity. On the basis of immunodiffusion tests, Parvez concluded that extracts of *Alcaligenes faecalis, Erwinia amylovora,* and *Klebsiella pneumoniae* were not releated to *Achromobacter nematophilus.*

Parvez also used the agglutination test to distinguish between bacteria (Table 176). The ability to inhibit agglutination was also used as a means of measuring relatedness between bacterial species (Table 177). A study of the immunofluorescence of the various strains also showed a degree of relativeness (Table 178.) The data obtained from immunodiffusion, immunoelectrophoresis, and agglutination studies showed few antigenic differences between *A. nematophilus* from the DD-136, Mexican, and Agriotos strains. These strains had not undergone as much antigenic divergence as the bacteria from *N. glaseri* and *N. bibionis.*

Tests conducted by Dr. Carl Bovelle and the present author showed that cells of *A. nematophilus* isolated from the DD-136 strain of *N. carpocapsae* were very fragile and susceptible to lysis in hypotonic solutions. The above bacterium was grown for 8 hr in

TABLE 173

Characteristics of the Relationship Between Steinernematidae and Heterorhabditidae and Their Associated Bacteria

Nematode family	Genera with associated bacteria	Location of bacteria in infective stage juveniles	Identification of bacteria	Diagnostic characters of the bacteria
Steinernematidae	*Steinernema*	Not known	*Flavobacterium* sp.	None reported
	Neoaplectana	Intestine	*Achromobacter* sp.	Catalase negative not bioluminescent
Heterorhabditidae	*Heterorhabditis*	Pharynx and intestine	Characterized, but not yet identified	Catalase positive bioluminescent

TABLE 174

Cultural and Biological Characteristics of the DD-136 and Agriotos strains of *Achromobacter nematophilus* isolated from Separate Populations of *Neoaplectana carpocapsae*[246]

Test (room temp = 24°C)	Bacterial strains[a]	
	DD-136 strain	Agriotos strain
Cell size (P-HOH; 24 hr)		
Rods (average length × width)	5.4 × 1.1 m	5.0 × 1.1 m
Spheres (average diameter)	2.5 m	2.4 m
Motility broth	+	+
Flagellation	Peritrichous	Peritrichous
Chromogenesis (NA and P-HOH)	−	−
Aerobic (AC medium)	+	+
Litmus milk (7 days)	+	+
Reduction peptonization	+	+
Curd	+	+
Loeffler's blood serum (5 days)		
Proteolysis	+	+
Pigment	Light brown	Light brown
Gelatin liquefaction (7 days)	+	+
MacConkey agar (24 hr)	+ w	+
Triple sugar iron agar	NC, NC, −	NC, NC, −
H₂S (TSI) (24 hr)	−	−
Potato slant (3 days)		
Growth	+	+
Pigment	−	−
Catalase	−	−
Oxidase	−	−
Cytochrome oxidase	−	−
Lysine decarboxylase	−	−
Arginine dihydrolase	−	−
Ornithine decarboxylase	−	−
Simmons citrate	−	−
Phenylalanine deaminase	−	−
Indole production	−	−
Nitrate reduction	−	−

[a] Symbols and abbreviations: −, negative; +, positive; + w, weakly positive; NC, no charge; BtB, bromothymol blue indicator; BCP, bromocresol purple indicator.

TABLE 174 (continued)

Cultural and Biological Characteristics of the DD-136 and Agriotos strains of *Achromobacter nematophilus* isolated from Separate Populations of *Neoaplectana carpocapsae*[246]

Test (room temp = 24°C)	Bacterial strains[a]	
	DD-136 strain	Agriotos strain
Methyl red	−	−
Voges-Proskauer	−	−
Hugh and Leifson's oxidation - fermentation medium		
Open tube	+	+
Closed tube	+	+
Anaerobic NA culture	+ w	+ w
Fermentations (24 hr)	BtBP	BtBP
Adonitol	+ w:−	+ :−
Arabinose	+ :−	+ :−
Glucose	+ : +	+ : +
Dulcitol	+ w: + w	+ w: + w
Esculin	−: + w	−: + w
Fructose	+ w: + w	+ w: + w
Inositol	+ w: + w	+ w: + w
Lactose	+ w: + w	+ w: + w
Mannitol	+ w: + w	+ w: + w
Melezitose	+ w: + w	+ w: + w
Melibiose	+ w: + w	+ w: + w
Raffinose	+ w: + w	+ w: + w
Saccharose	+ w: + w	+ w: + w
Salicin	+ w: + w	+ w: + w
Sorbose	+ w: + w	+ w: + w
Sucrose	+ w: + w	+ w: + w
Trehalose	+ : + w	+ : + w
Xylose	+ w: + w	+ w: + w

a 1% peptone, 0.5% NaCl, and 0.5% glucose liquid culture, centrifuged, and resuspended in tris-buffer for polarographic measurements of oxygen consumption. The resting cell suspension was capable of respiring malate (10mM sodium malate), aspartate (10 mM sodium aspartate), succinate (10 mM), glutamate (10 mM), and glucose (10 mM).

The general conclusions after examining the results of these tests, along with the original characterization of *A. nematophilus,* were that this bacterium does not fit in any of the existing genera and probably warrants a separate taxon.

Similar conclusions were made regarding the status of bacteria associated with various species of *Heterorhabditis*. These bacteria resemble *A. nematophilus,* biologically and culturally, but differ in several important characters. Cultural and biological characteristics of the bacterial strains isolated from *H. bacteriophora* and *H. heliothidis* are presented in Table 179.

II. INQUILINISM

This type of symbiosis is characterized by the simple presence of a microorganism in or on another organism, without mutual benefit or harm. This section discusses the

TABLE 175

Growth Studies of the Agriotos and DD-136 Strains of *Achromobacter nematophilus* Under Varying Conditions of Temperature, pH, and Salt Concentration[246]

Conditions studied		Absorption of bacterial strains[a]		
		DD-136 strain	Agriotos strain	Control
Temperature tolerance	°C			
(pH 7,	10	1	1	0
0.5% NaCl)	24	6	5	0
	30	14	10	0
	37	1	1	0
	42	0	0	0
pH tolerance	pH			
(24°C,	5.0	0	0	0
0.5% NaCl)	6.0	0	0	0
	7.1	69	4	0
	8.2	57	2	0
	9.0	31	3	0
	10.0	0	0	0
Sodium chloride	% conc			
tolerance	of NaCl			
(24°C,	0.0	0	0	0
pH 7)	0.5	90	67	0
	1.0	82	62	0
	3.0	1	1	0
	4.0	0	1	0
	5.0	0	1	0

[a]   Growth expressed as percent absorbance in 24 hr P-HOH culture, measured in a Bausch and Lomb Spectronic 20 colorimeter-spectrophotometer, at a wavelength of 600 nm. A reading of 1 on the Spectronic 20 is considered equivalent to 0 for the purposes of this study.

subject of microorganisms being carried into insect hosts by attachment on the cuticle of infective stage nematodes. Such an occurrence can be very detrimental for the nematode since most insect parasitic nematodes require a clean, uncontaminated host for successful development. Microorganisms being introduced into the hemocoel of healthy insects could multiply and destroy the host, thereby eliminating the parasite.

Only members of the Steinernematidae and Heterorhabditidae habitually carry specific microorganisms into the body cavity of their hosts. However, all nematodes that enter hosts directly from the environment have the potential of bringing microorganisms into the insect.

Even infective juveniles of *Neoaplectana carpocapsae* may carry bacteria on their cuticles as shown by scanning electron microscopy (Figure 67). These foreign bacteria, which may occur in the digestive tract of the insect, then compete with the symbiotic bacteria in the host, and nematode development depends very much on the ratio of "good" and "bad" bacteria. When the host blood principally contains the symbiotic bacterium, nematode reproduction is generally high; however, when foreign bacteria get the upper hand, nematode development may be halted. This is why it is so important to have "clean" infective stage juveniles.

The case with mermithid nematodes is interesting. There is no known mermithid whose development is dependent on a microorganism growing alongside it in the body cavity of a host. In fact, sterile conditions are essential for successful development. Thus, do the infective stages of mermithid nematodes carry microorganisms into the

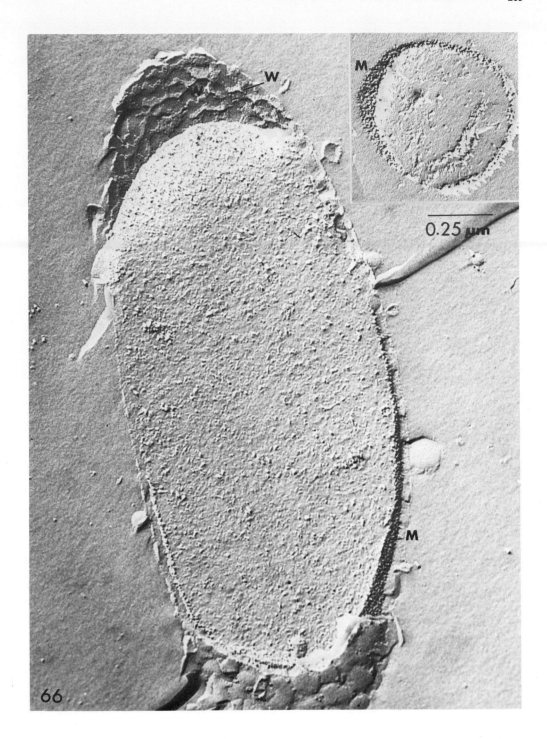

FIGURE 66. Freeze-etch electron micrograph photograph of a cell of *Achromobacter nematophilus* showing outer wall (W) and inner membrane (M). Insert shows the same of a protoplast with only the membrane (M) present. (Courtesy of G. De Zoeten.)

host? If so, then the host's defense system would have to cope with them. If microorganisms are not carried into the host, is this because they don't stick to the cuticle or because they are wiped off during penetration? In order to partially answer these ques-

TABLE 176

Agglutination Titer of *Achromobacter nematophilus* and Other Bacterial Strains Tested Against DD-136 Antiserum[345]

Organisms (1 × 10⁶ cells/m*l*)	Isolated from	Agglutination titer
Achromobacter nematophilus (homologous strain)	*Neoaplectana carpocapsae* DD-136 strain	1/800
A. nematophilus	*N. carpocapsae* Mexican strain	1/200
A. nematophilus	*N. carpocapsae* Agriotos strain	1/400
Achromobacter sp.	*N. glaseri*	—
Achromobacter sp.	*N. bibionis*	—
Alcaligenes faecalis	ATCC 8750	—
Klebsiella pneumoniae	MIA5	—
Erwinia amylovora	FB-128	—

Note: Titers with antiserum dilutions below 100 were regarded as negative.

TABLE 177

Agglutination Inhibition of *Achromobacter nematophilus* DD-136 Strain by the Cell-free Extracts from Heterologous Strains[345]

Bacterium	Isolated from	Antiserum	Serum dilutions			
			1/100	1/200	1/400	1/800
Achromobacter nematophilus	*Neoaplectana carpocapsae* DD-136 strain	H-DD-136	+	+	+	+
A. nematophilus	*N. carpocapsae* Agriotos strain	H-DD-136	+	+	−	−
A. nematophilus	*N. carpocapsae* Mexican strain	H-DD-136	+	−	−	−
Achromobacter sp.	*N. glaseri*	H-DD-136	−	−	−	−
Achromobacter sp.	*N. bibionis*	H-DD-136	−	−	−	−
Alcaligenes faecalis	ATCC 8750	H-DD-136	−	−	−	−
Klebsiella pneumoniae	MIAS	H-DD-136	−	−	−	−
Erwinia amylovora	FB-128	H-dd-136	−	−	−	−

tions, the present author placed the infective stage juveniles of *Romanomermis culicivorax* in sterile water containing heavy suspensions of the bacteria *Pseudomonas aeruginosa* and *Serratia marcescens*. After 30 min, the nematodes were removed by centrifugation, washed three times, centrifuged, and examined with a scanning electron microscope. It was noted that whereas no cells of *P. aeruginosa* could be found on the surface of washed infective stage juveniles, rods of *S. marcescens* were clearly visible (Figure 68). This suggests that some bacteria, including those that are capable of causing insect disease, may not be attracted to nematode cuticle, whereas others, also considered potential insect pathogens, are definitely attached. Whether the latter microorganisms would be removed by the insect's cuticle as the nematode penetrated is not known.

TABLE 178

Fluorescent Antibody Staining Reactions of *Achromobacter nematophilus* Strains and Other Bacteria Using DD-136 Antiserum[345]

Bacterium[a]	Isolated from	Direct test[b]	Indirect test	FA titer
Achromobacter ne-matophilus	*Neoaplectana carpocapsae* DD-136 strain	4,*	4 +	1/100
A. nematophilus	*N. carpocapsae* Mexican strain	3 +	3 +	1/40
A. nematophilus	*N. carpocapsae* Agriotos strain	4 +	4 +	1/80
Achromobacter sp.	*N. glaseri*	—	—	—
Achromobacter sp.	*N. bibionis*	—	—	—
Alcaligenes	*faecalis*	—	—	—
Klebsiella	*pneumoniae*	—	—	—
Erwinia	*amylovora*	—	—	—

[a] 10^6 cells per milliliter.
[b] Intensity of fluorescence.

TABLE 179

Cultural and Biological Characteristics of the Bacteria Associated with *Heterorhabditis bacteriophora*[315] and *H. heliothidis*[316]

Test (24°C)	*H. bacteriophora*	*H. heliothidis*
Cell size (μm) (n = 25)		
Rods (length)	5.2 (3.1—9.8)	6.0 (2.0—7.8)
Rods (width)	1.3 (1.2—1.7)	0.8 (0.6—1.1)
Spheres (diameter)	2.7 (2.0—3.7)	
Motility	+	+
Flagellation	Peritrichous, close coiled	Peritrichous
Chromogenesis (NA and P-HOH)	Yellow-brown	Pink-red
Litmus milk (7 days)		
Reduction peptonization	+	—
Curd	—	—
Loeffler's blood serum (5 days)		
Proteolysis	w	+
Pigment	Yellow-orange	NT
Gelatin liquefaction (7 days)	—	+
MacConkey agar (24 hr)	w	NT
Triple sugar iron agar	NC, NC, −	—
H_2S (TSI) (24 hr)	—	—
Aerobic (AC) medium	+	NT
$NO_3 \rightarrow NO_2$	—	—
Potato slant (3 days)		
Growth	+	NT
Pigment	Yellow	NT
Catalase	+	+
Oxidase	—	—
Cytochrome oxidase	—	NT
Lysine decaroxylase	—	—
Arginine dihydrolase	—	—
Ornithine decarboxylase	—	—

TABLE 179 (continued)

Cultural and Biological Characteristics of the Bacteria Associated with *Heterorhabditis bacteriophora*[315] and *H. heliothidis*[316]

Test (24°C)	*H. bacteriophora*	*H. heliothidis*	
Gram reaction	−	−	
Bioluminescence	+	+	
Simmons citrate	−	−	
Phenylalanine deaminase	−	−	
Indole production	−	−	
Methyl red	−	−	
Voges-Proskauer	−	−	
Hugh and Leifson's			
O-F medium			
Open tube	+	w	
Closed tube	+	+	
Fermentations	BtB[a]	BCP[b]	
Adonitol	w	−	vw
Arabinose	+	vw	+
Glucose	+	+	+
Culcitol	+	vw	w
Esculin	pl	w	NT
Fructose	+	vw	+
Inositol	+	vw	vw
Lactose	+	vw	+
Mannitol	+	vw	w
Melezitose	+	vw	w
Melibiose	+	vw	w
Raffinose	+	vw	w
Rhamnose	+	vw	+
Saccharose	+	vw	NT
Salicin	+	vw	w
Sorbose	+	vw	+
Sucrose	+	vw	+
Trehalose	+	vw	+
Xylose	+	vw	+

Note: −, negative; +, positive; w, weak; vw, very weak; NC, no charge; NT, not tested.

[a]   Bromothymol blue.
[b]   Bromocresol purple.

FIGURE 67 and 68.   (67) Bacterial cells attached to the cuticle of an infective juvenile of *Neoaplectana carpocapsae*. (Courtesy of Dawn Hammond). (68) Cells of *Serratia marcescens* attached to the cuticle of a preparasitic juvenile of *Romanomermis culicivorax*. (Courtesy of Dawn Hammond.)

Chapter 6

IMMUNITY TO ENTOMOGENOUS NEMATODES

I. INTRODUCTION

The fact that most entomogenous nematodes are fairly host specific has already been established, since many attack hosts belonging to a single genus or family. There are many reasons why insect parasitic nematodes cannot attack other hosts in the same habitat as the normal host, and some of these are discussed below. The various categories of immunity mentioned here are host escape, cellular responses, and humoral responses. Earlier cases of insect immunity to parasitic nematodes will not be reported since they have been presented earlier.[346,347]

II. HOST ESCAPE

Most insects avoid or escape nematode parasitism by their normal habits or behavior or by the structure of their cuticle or conditions inside their digestive tracts. Recently, a study showed how this type of immunity can develop in an insect population that was originally completely susceptible to a specific parasite.[76]

Larvae of *Anopheles quadrimaculatus* had been used during 4 years for culturing the mermithid *Strelkovimermis peterseni* at the Gulf Coast Mosquito Research Laboratory in Lake Charles, Louisiana. During this period a decline in nematode production was noticed. Since the normal rearing procedure was to return host pupae that had survived infection back to the mosquito colony for adult production, the question of host immunity was raised. When larvae of *A. quadrimaculatus* which had experienced no previous contact to the parasite were used, they produced twice as many infected hosts and twice the average number of nematodes than the original host. The authors concluded that by restocking their colony with individuals that escaped infection, a type of immunity was bred into the population. They noticed that host larvae of the strain originally selected were very active and tried to bite the attacking nematodes, whereas those of the unselected strain were more quiescent. The nature of the resistance was considered a type of host escape through increased host activity bred into the population.

III. CELLULAR RESPONSES

The most common type of cellular response exhibited by insects against metazoan parasites is encapsulation.

Simple encapsulation is an accumulation of host cells around a parasite without any secondary melanization reactions. It is illustrated in Figure 69 showing an adult of *Mesodiplogaster lheritieri* within a hemocytic capsule of *Galleria mellonella* larvae. Multiple layers of blood cells are deposited around the host to form a spherical capsule. The enclosed nematodes usually die 3 or 4 days following this reaction.

Rarely, mosquitoes may encapsulate parasites. An example of this is the response of *Culex terratans* to the developing parasitic juveniles of the mermithid *Romanomermis culicivorax*. Parasites 2 or 3 days old are covered with host blood cells and fail to complete their development (Figure 70). Such reactions are important to note in biological control studies since they may alter expected results.

More frequently found is the condition known as melanotic encapsulation, or when melanization accompanies the encapsulation response. A classic example of this type

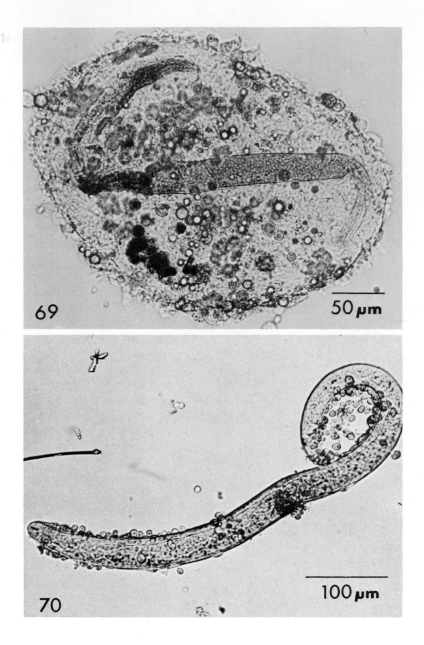

FIGURE 69 and 70.   (69) *Mesodiplogaster lheritieri* enclosed in a capsule of blood cells from *Galleria mellonella*. (70) A parasitic juvenile of *Romanomermis culicivorax* with attached blood cells from a *Culex terratans* larva.

was reported in *Diabrotica* beetle larvae attacked by the mermithid *Filipjevimermis leipsandra.* Host hemocytes lysed on the cuticle of the nematode soon after it entered the host, and within 6 to 8 hr after entry, an inner layer of melanin had formed around the nematode (Figures 71 and 73).

This type of host response was also reported in fly larvae attacked by *Heterotylen-chus autumnalis*[328] and in certain species of mosquitoes parasitized by *R. culicivorax.*[39] Thus, larvae of *Psorophora confinnis, Culex territans,* and *Aedes triseriatus* exhibited various degrees of melanotic encapsulation against developing *R. culicivorax,* and parasitic development was never completed in the latter two hosts. Just what enables these

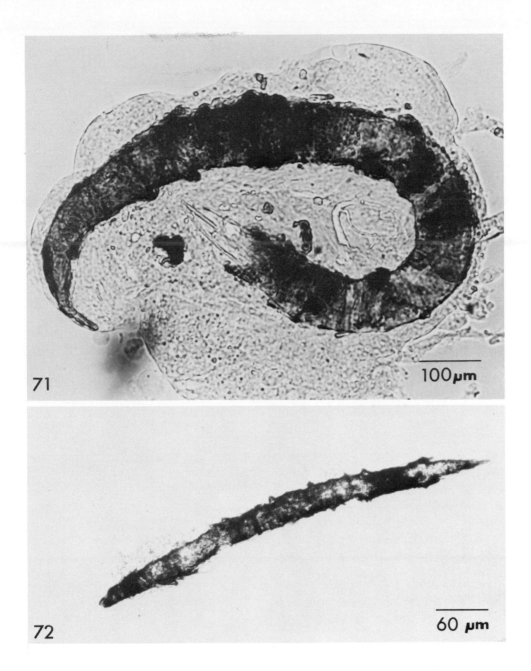

FIGURE 71 and 72. (71) A juvenile of *Filipjevimermis leipsandra* enclosed in a melanotic capsule formed in a *Diabrotica* larva. (72) Infective juvenile of *Neoaplectana carpocapsae* melanized in a *Culex pipiens* larva.

mosquitoes to elicit a response when development can proceed normally in more than 15 other mosquitoes species is difficult to answer. A similar host reaction with the same parasite was reported in *Anopheles sinensis*, a common mosquito in rice fields in southeast Asia. Mitchell et al.[23] noted that soon after invasion (within 24 hr) the nematodes became melanized and failed to complete their development. It is not known whether some of the above reports are a case of melanotic encapsulation or humoral melanization.

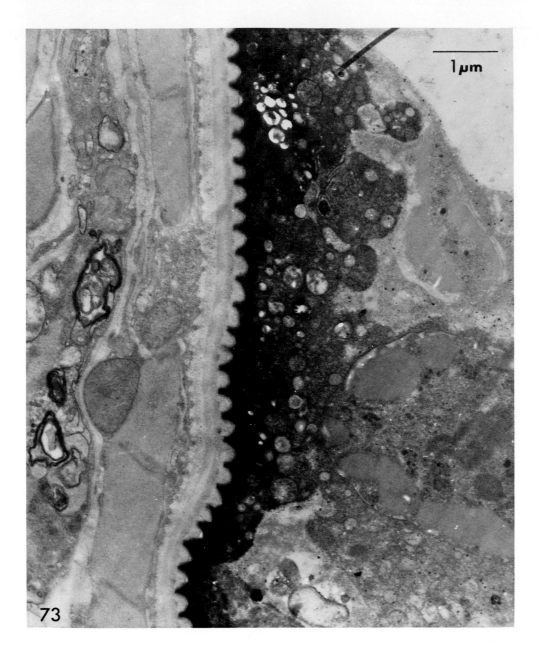

FIGURE 73.   Electron micrograph showing a layer of melanin adjacent to the cuticle of *Filipjevimermis leipsandra* removed from a *Diabrotica* larva.

IV. HUMORAL RESPONSES

In this category are grouped the noncellular components of the insect hemolymph which have an adverse effect on nematode development. There are cases of a nematode entering a host, but not growing and eventually perishing without any outward signs of attack. In such cases, there are probably substances in the host's hemolymph that are either toxic to the parasites or the absence of substances needed for the completion of parasitic development.

Another response in this category is humoral melanization, or melanization by noncellular components of the blood. This type of response was described in *Culex pipiens* larvae invaded by infective stage juveniles of *Neoaplectana carpocapsae*[348] (Figure 72). Soon after entry into the host's hemocoel, a homogenous deposit surrounded the nematode. Pigment granules began to form within this deposit, and they eventually enlarged and coalesced to form a layer of melanin around the parasite. The original deposit was formed by components in the noncellular portion of the host's hemolymph. Other cases of humoral melanization have been reported in midges.[349] In this case, melanin was deposited directly on the surface of the nematode without the direct participation of blood cells. The present author observed an example of humoral melanization in midges of the genus *Chironomus* that were attacked by the preparasitic juveniles of *R. culicivorax*. Soon after penetration, the newly entered juveniles were encased in a solid layer of pigmented material, presumed to be melanin. When newly hatched midges were exposed to numerous preparasiteic juveniles, they usually were killed as a result of mass penetration. Parasitic development in this host was not observed.

Another type of humoral response may occur in blackflies attacked by the mosquito mermithid *R. culicivorax*. It has been shown that the preparasites of this nematode will enter and initiate development in first, second, and third stage larvae of several blackfly species. Surprisingly, there is no direct host reaction against this foreign parasite, but that may be because blackflies in general do not have the ability to rapidly encapsulate or melanize foreign parasites. Eventually both nematode and host die and, to date, no one has been able to keep the blackflies alive long enough to determine if the parasite can complete its development in this "abnormal" host. This may be due to the problems involved in culturing infected blackflies under laboratory conditions, or it may be due to a substance given off by the parasite that is toxic to the host or vice versa. If normal development of *R. culicivorax* could be completed in any blackfly species, it would be an unusual event.

There are other unusual types of insect response that occur only rarely against nematode parasites. These include intracellular melanization and responses of certain tissues other than blood cells, such as intestinal epithelium and fat body cells.[347] These responses have been noted against the smaller insect-parasitic stages of vertebrate nematodes belonging to the Spirurida. Of course, the other host responses discussed above may also be elicited by these parasites as well.

Thus, from the standpoint of biological control, it is important to test host populations against the parasite before actual applications are made in order to determine if any host immunity is present. On the other hand, the specificity of entomogenous nematodes can be an advantage itself, since beneficial insects in the same habitat as the target host will not be parasitized.

Chapter 7

NATURAL ENEMIES OF ENTOMOGENOUS NEMATODES

I. INTRODUCTION

The natural enemies of entomogenous nematodes, as well as those of nematodes in general, can be grouped into predators, parasites, and pathogens. The effect of these agents on natural populations of entomogenous nematodes is little known, but may be considerable at times. However, these agents are not only effective in nature, but also in the laboratory when attempts are being made to culture entomogenous nematodes. For example, mermithid nematodes are especially susceptible to fungal infections under laboratory conditions, and such infections may seriously hamper mass production of these parasites. The various agents are discussed below under their respective headings.

II. PREDATORS

A number of invertebrates prey on nematodes in general[350,351] and will undoubtedly attack the free-living stages of many entomogenous nematodes, depending on their size. Some predators of insect nematodes have been noted (Figure 74). One of these is the crustacean *Cyprinotus dentatis* Sharpe. This ostracod was observed to prey upon the infective stage juveniles of the mermithid *Romanomermis culicivorax* in Taiwan.[23]

In laboratory experiments with *R. culicivorax* infecting mosquito larvae, the parasitism of mosquitoes was reduced when ostracods were present, showing that a significant number of infective stage nematodes were removed from the infection chamber. Ostracods are common in warm waters rich in organic matter and may pose a problem when mermithid nematodes are added to such habitats.

Another common group of nematode predators are mites, especially members of the genus *Macrochiles*. Anyone culturing free-living nematodes on agar plates in the laboratory can testify to the detrimental results when mites find their way into the plates. The present author has observed mites feeding on the free-living infective stage juveniles of *Neoaplectana carpocapsae* in laboratory soil cultures, and these predators undoubtedly do the same in nature when the opportunity arises.

Actually, most predaceous arthropods are probably capable of attacking various stages of entomogenous nematodes. The present author noted an adult carabid beetle feeding on a postparasitic juvenile of *Hexamermis arvalis,* and Platzer and MacKenzie-Graham[352] recorded several predators attacking *R. culicivorax* in the laboratory.

The copepod (*Cyclops vernalis*), ostracods, the cladoceran (*Simocephalus vetulus*), and the gammarid (*Hyallela azteca*) attacked and devoured the preparasitic juveniles of this mermithid. The larger postparasitic juvenile nematodes, which are free-living, were attacked by the diving beetle (*Laccophilus terminalus*), the gammarid (*H. azteca*), dragonfly nymphs, damselfly naiads, isopods (*Asellus* sp.), and young crayfish (*Procambarus clarki*). These observations demonstrate the importance of potential predators on nematodes introduced into new localities.[352]

III. PARASITES

Parasites of entomogenous nematodes are not at all common in the literature. Steiner[353] mentioned the presence of two individuals of *Cephalobus* sp. inside a living *Agamermis paradecaudata* Steiner collected in Indonesia. The *Cephalobus* sp. deposited eggs inside the mermithid, which hatched after the latter died. Such a situation

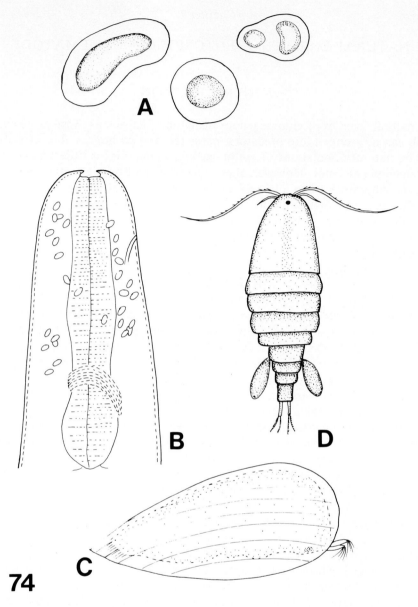

FIGURE 74.   Some enemies of entomogenous nematodes. (A) *Actinomyxidia* from *Agamermis paradecaudata.*[353] (B) The microsporidan *Nosema mesnili* in *Neoaplectana carpocapsae.*[354] (C) An ostracod. (D) A *Cyclops.*

must be rare, however, since cephalobids are not known to be parasitic on invertebrates and are considered free-living microphagous forms.

IV. PATHOGENS

A. Protozoa

Of the many groups of protozoa that attack invertebrates, only a few have been noted in insect nematodes. Steiner[353] reported a species of *Actinomyxidia* in the body cavity of *Agamermis paradecaudata* from Indonesia. The parasites were round to oval in shape and possessed a thick wall (Figure 74A). Their effect on the nematode was not known.

Veremchuk and Issi[354] discovered that when the Agriotos strain of the entomogenous nematode *Neoaplectana carpocapsae* parasitized lepidopterous hosts earlier infected with microsporidans, the nematodes also acquired the infection. Thus, when the caterpillars of *Pieris brassicae* L. and *Agrotis segetum* Schiff were infected with *Nosema mesnili* and *Plistophora schubergi*, respectively, and then attacked by *N. carpocapsae,* the developing nematodes were infected with both microsporidan species (Figure 74B).

B. Bacteria

There are few true bacterial infections of nematodes reported in the literature and certainly none of insect nematodes that have been proven to be pathogenic. Bacteria often cause problems in the storage of entomogenous nematodes in water by depleting the available supplies of oxygen.

Swain et al.[202] mentioned a possible bacterial infection in stored *Neoaplectana* nematodes. They reported that the infected nematodes were misshapen, apparently because of large masses of bacteria stored in their body cavity. Such affected nematodes never survived well in artificial culture and eventually died.

Another bacterialike organism was found causing a cuticular infection of *Thelastoma pterygoton* living in the larvae of the rhinocerous beetle *Oryctes monoceros* in West Africa[355] (Figure 75). The microorganisms were closely appressed to the nematode's surface and appeared to dissolve portions of the nematode's cuticle. Nematodes so infected showed an irregular surface since portions of the body were partially constricted from the remainder. It is possible that nematode mortality could occur when the cuticle became very thin and exposed the underlying hypodermis. The microorganisms were irregular in shape and size and contained a well-defined cell wall.

C. Viruses

Virus-like particles have been recovered from several nematodes, including the mermithid *Romanomermis culicivorax*[356] (Figure 76). In the latter case, isometric, enveloped particles measuring 50 nm in diameter were found in the hypodermal cords, nerve cells, and pseudocoelom of the infective stage juveniles. Although damage to the hypodermal cords was present, it is not known what effect these particles have on the well-being of the nematodes.

D. Fungi

A wide range of fungi attack nematodes and the literature on nematophagous fungi is very extensive (Figure 77). In 1925, Steiner[353] noted that fungi belonging to the Saprolegniaceae and Chytridinae were frequently found on mermithids and could be detrimental to a rearing or breeding program. Fungi do constitute a problem with mermithid nematodes and can destroy whole colonies. The present author has found mermithids in nature infected with fungi and noted that many of these individuals appeared to be injured. Since it is possible for one portion of a mermithid to be infected with pathogens and the other portion still moving, the question arises whether many of the fungi that have attacked the mermithid at injured locations are saprophytes. Bringing mermithids into the laboratory establishes artificial conditions that may be very stressful to the nematodes and make them more susceptible to fungal invasion. Thus, some fungal outbreaks may indicate that conditions are suboptimum.

Nolan[357] recorded *Saprolegnia megasperma* Coher on a field-collected specimen of the blackfly mermithid *Mesomermis flumenalis*. The sign of the infection at the time of collection was hyphal tips growing out of the nematode's body. He noted that since this nematode is being considered for future field releases, *S. megasperma* and other microorganisms might affect the degree of success.

Fungi of the genus *Catenaria* also attack mermithid nematodes (Figure 78). Nolan[357]

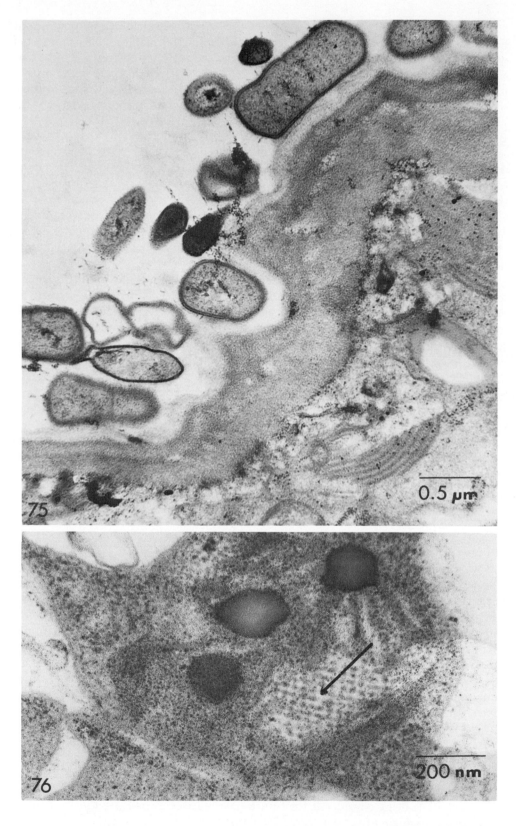

FIGURE 75 and 76.   (75) Bacteria-like microorganisms digesting the cuticle of *Thelastoma pterygoton*.   (76) Virus-like particles (arrow) in the hypodermal cord of a preparasitic juvenile of *Romanomermis culicivorax*.

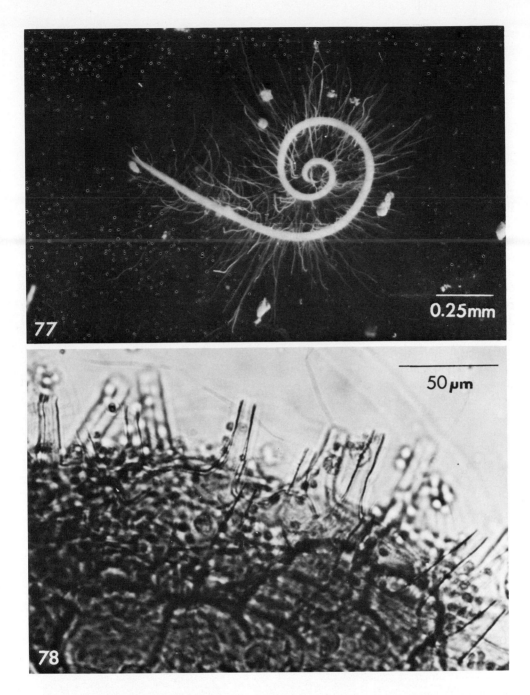

FIGURE 77 and 78.   (77) A *Gastromermis* sp. from California killed by a fungus. (78) The exit tubes of *Catenaria* emerging from the body of a *Gastromermis* sp. from California.

observed that *C. anguillulae* Sorokine could attack the eggs of *Romanomermis culici-vorax*. Laboratory epizootics of *C. anguillulae* also occurred in culture of the same nematode.[357a] Up to 90% of the postparasites were infected, thus greatly reducing the potential numbers of infective stage juveniles expected. Infection was initiated by motile zoospores discharged from the sporangia of *C. anguillulae*. The zoospores would attach themselves to the nematode's cuticle and then enter the body cavity of the host. Eventually zoosporangia and nesting sporangia were formed inside the nematode. All stages of *R. culicivorax,* including the eggs, were susceptible to infection, and the fungus was also capable of growing saprophytically in dead nematodes.

The disease was almost completely eradicated by adjusting the pH of the water to 4.5 with acetic acid. Low pH prevents zoospore motility and germination, yet has no effect on the nematodes.

V. STRUCTURES CONFUSED WITH PATHOGENS

A. Protein Platelets

Protein platelets found in the pseudocoelom of some mermithid nematodes have been categorized under the heading of "enigmatic parasites" and "blood globules." Protein platelets from a mermithid belonging to the genus *Gastromermis* (earlier identified as *Hydromermis* sp.) were examined by Poinar et al.[358] (Figure 79). These platelets had an average diameter of 23.2 μm (14.6 to 30.5) and an average width of 5.5 μm (3.7 to 7.3). They were hyaline in appearance, lacked a nucleus, and contained an indistinct limiting membrane. Some were broken, but all floated freely in the pseudocoelom, being especially visible in the tail region. Under the electron microscope, the platelets were homogeneous and possessed a structure similar to a crystalline lattice. The platelets gave a strong positive reaction for protein; however, their function, if any, could not be determined. It is probable that they represent a nontoxic method of storing accumulated metabolites produced during nematode growth.

Such platelets are fairly common in some species of mermithids and, indeed, provide a means of identification in postparasitic juveniles. A smaller type of protein platelet occurs in the body cavity of *Romanomermis culicivorax* (Figure 80.)

FIGURE 79   Protein platelets in the pseudocoelom of a *Gastromermis* sp.

FIGURE 80.   Ultrastructure of a protein platelet from the pseudocoelom of *Romanomermis culi-civorax*.

Chapter 8

ENVIRONMENTAL IMPACT OF ENTOMOGENOUS NEMATODES

I. INTRODUCTION

With the widespread distribution of entomogenous nematodes for biological control, there is always concern of possible effects on nontarget organisms such as plants, other invertebrates, and vertebrates, including man. Fortunately, many insect-parasitic nematodes have a narrow host range, attacking members of a single genus or family. However, the neoaplectanids and heterorhabditids are able to attack insects from several orders; thus, more caution must be exercised if beneficial insects are involved. The known effect of entomogenous nematodes on the various groups of nontarget organisms will be discussed below.

II. POSSIBLE EFFECT ON PLANTS

Some nematodes parasitizing wood-boring insects possess free-living stages which feed on fungi. Members of the genus *Deladenus* have gone a step further and possess free-living cycles that sustain themselves on fungi symbiotically associated with wood wasp hosts. The species of fungi involved (members of the genus *Amylostereum*) are considered harmful in many cases, and therefore their possible reduction by feeding nematodes would not be significant. Another species of *Deladenus* attacking barine weevils has a free-living cycle on *Brassica* plants; however, damage to the plant was not reported and is probably insignificant.

Members of the genus *Fergusobia* attack *Eucalyptus* gall flies and have a free-living generation which feeds on the plant gall cells. Again, the plant-feeding stage is not considered a pest to the *Eucalyptus* tree since it apparently only survives within the leaf gall formed by the stimulus of the fly.

Aside from the above-mentioned two genera in the Neotylenchidae and some of the free-living stages of the aphelenchoidids, no other entomogenous nematodes are known to feed on fungi or vascular plants.

III. POSSIBLE EFFECT ON NONTARGET INVERTEBRATES

We know that neoaplectanid and heterorhabditid nematodes have a wide host range and potentially could infect a number of insects in a treated area. However, most noninsect invertebrates appear to be resistant to these nematodes as demonstrated by preliminary infection attempts cited in Table 180.

Since many stream invertebrates are important entities in the food chain for fish and other vertebrates, some tests have been conducted with the mermithid *Romanomermis culicivorax* to determine if nontarget invertebrates are susceptible. Thus, Ignoffo et al.[38] and Otieno[36] exposed a number of nonmosquito hosts to the preparasitic juveniles of *R. culicivorax*. The results are presented in Table 181.

None of the invertebrates tested were able to support the complete development of *R. culicivorax,* and on the basis of these tests, one can assume that the widespread distribution of this mermithid will not affect organisms other than mosquitoes.

IV. POSSIBLE EFFECT ON VERTEBRATES OTHER THAN MAN

Thus far, only two species of entomogenous nematodes have been tested against vertebrates, e. g., *R. culicivorax* and *Neoaplectana carpocapsae.* Preparasitic juveniles

TABLE 180

Effects of *Neoaplectana carpocapsae* on Some Invertebrates
Other Than Insects

Invertebrate			
Group	Species	Results	Ref.
Acari	*Tyrophagus noxius* Zachvatkin	—	163
Symphyla	*Scutigerella immaculata* (Newport)	+	360
Diplopoda	Garden milliped	—	258
Mollusca	*Deroceras reticulatum* (Mull.)	—	143
	Garden slug	—	258
Annelida	*Aporrectodea* sp.	—	143
Crustacea	*Cambarus* sp.	—	143
	pill bug	—	258

TABLE 181

Results of Exposure of Some Invertebrates to Preparasitic Juveniles of *Romanomermis culicivorax* (Exposed to Approximately 1000 Nematodes)

Invertebrate	Results	Ref.	Invertebrate	Results	Ref.
Crustacea			Coleoptera		
Cambarus sp.	—	36	*Dytiscidae* (undetermined)	—, 1	38
Daphnia magna Straus	—	38	*Laccophilus* sp.	—	36
Cyclops sp.	—	38	*Haliplidae* (undetermined)	—	38
Annelida			*Hydrophilidae* (undetermined)	—, 1	38
Lumbriculus sp.	—	38			
Insecta			*Hydrophilus triangularis* Say	—	36
Ephemeroptera			*Helophorus* sp.	—	36
Argia sp.	—	36	*Tropisternus lateralis* (Fab.)	—	36
Baetedae (undetermined)	—	38	*Gyrinus punctellus* Ochs	—	36
Callibaetis sp.	—	36	Diptera		
Odonata			*Corethrella appendiculata* Grab.	1	38
Zygoptera (undetermined)	—	38	*Corethrella brakeleyi* (Coq.)	1	38
Hemiptera			*Chironomus* spp.	2	38
Belostomatidae (undetermined)	—	38	*Chironomus* spp.	2	36
Corisella decolor (Uhler)	—	36	*Tanyper* spp.	2	36
Corixidae (undetermined)	—	38	*Psychoda* sp.	—	38
Notonectidae (undetermined)	—	38	*Simulium damnosum* Theo.	2	29
Notonecta unifasciata Guerin	—	36	Simulium spp.	1	40

Note: — = no penetration; 1 = penetration, but no development; 2 = penetration and initial development only.

of the former species were exposed to fish, tadpoles, mice, and rats by Ignoffo et al.[38,359] and Otieno,[36] respectively (Table 182). No sign of infection was noted, even when the mice and rats were exposed to intranasal, intraperitoneal, per os, and dermally administered nematodes. There also was no sign of weight loss and no difference in tissue appearance of the treated mammals. Some dead nematodes were recovered from feces of treated mice.[359]

TABLE 182

Results of Exposure of Some Vertebrates to the Preparasitic Juveniles of *Romanomermis culicivorax*

Vertebrate	Results	Ref.
Fish		
Gambusia affinis (B. and G.)	—	36
Ictalurus punctalus (Raf.)	—	38
Micropterus salmoides (Lac.)	—	38
Pimephales promelas (Raf.)	—	38
Salmo gairdneri Rich.	—	38
Amphibians		
Bufo boreas halophilus (tadpole stage)	—	36
Mammals		
White albino mice	—	359
White rats	—	359

Another nematode that has been mass produced, *Neoaplectana carpocapsae,* has also been tested against vertebrates.[359] All stages of the above species were fed to laboratory rats, and no effect was observed. Those nematodes recovered from feces of treated animals were dead, and autopsies showed no inflammation of the alimentary tract or establishment of the nematode in that location.

Similar negative results were also obtained by Gaugler[359b] when he introduced infective stage *N. carpocapsae* into rats (per os and intraperitoneal inoculation). Nematodes recovered from intraperitoneal inoculated rats were usually coated by vertebrate macrophages (Figure 81). Sometimes groups of nematodes were stuck together in clusters, all coated with host cells (Figure 82). Thus, the nematodes definitely elicit a host response in the vertebrate and apparently die without showing any signs of development.

V. ASSOCIATIONS OF ENTOMOGENOUS NEMATODES WITH MAN

Of the wide range of entomogenous nematodes, only members of the family Mermithidae have occasionally been reported as accidental parasites of man. These reports are disturbing since certain representatives of this family have been and are being distributed for insect control. The following is a review of all such cases in the literature and presentation of two new cases.

The present author received specimens and assistance for this study from the following: S. Prudhoe, Department of Zoology, British Museum of Natural History, London; L. A. Leon, Laboratorio de Parasitologia y Medicina Tropical, Quito, Ecuador; W. R. Nickle, Plant Nematology Laboratory, Beltsville, Maryland; A. Morgan Golden, Plant Nematology Laboratory, Beltsville; and Gordon W. Vacura, curator, Medical Museum, Washington, D. C.

A. Case 1: *Agamomermis hominis oris* (Leidy) (syn. *Filaria hominis oris* Leidy 1850)

Location — Described from a single preserved specimen labeled "obtained from the mouth of a child."[361]

Description — Body thread-like, white, opaque; length, 140 mm; greatest width, 0.38 mm; head round; width at head, 0.101 mm; tail obtuse; width, 0.317 mm; tail bearing a spike 0.050 mm long and 0.013 mm wide at base.

Comments — Although the type specimen has been lost or misplaced and is not available for study,[361a] (Stiles[362] indicated that it probably had been destroyed), the

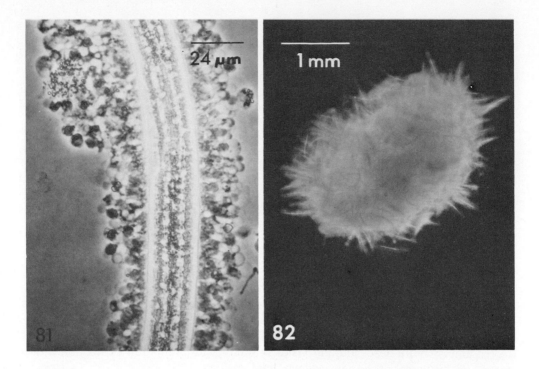

FIGURE 81 and 82. (81) Rat macrophages surrounding an infective juvenile of *Neoaplectana carpocapsae.* (Courtesy of R. Gaugler.) (82) A cluster of *Neoaplectana carpocapsae* infective juveniles stuck together with macrophages originating from a rat. (Courtesy of R. Gaugler.)

description fits that of a mermithid. Also, Leidy[363] later identified some mermithids collected from an apple as *Mermis* (syn. *Hexamermis*) *acuminata* and stated that they could have been the same species as *A. hominis oris.* He also collected a similar mermithid from codling moth larvae, which explains how a child might have encountered the parasite. Stiles[362] also reported finding mermithids from apples. Thus, although we cannot definitely reconstruct the circumstances surrounding the presence of a mermithid in a child's mouth, it is possible that it came from an insect that was ingested by the child directly or indirectly (through eating infested fruit).

B. Case 2: *Agamommermis restiformis* (Leidy) (syn. *Filaria restiformis* Leidy[364])

Location — Said to have been removed from the urethra of a man in the U.S.

Description[364] **(Figures 90 and 91)** — Vivid red in color and very active when collected; 66 cm long, greatest width 1.5 mm; caudal end curved, bluntly rounded; lacking an anus or genital aperture; mouth a terminal pore without lips, papillae or any kind of armature; pharynx cylindrical, opening into a straight cylindrical intestine which apparently ends in a blind pouch; body width at neck region 0.375 mm; width at beginning of esophagous 1.125 mm. The following supplementary description is taken from Stiles[365] (Figures 89 and 92). Cuticle composed of several concentric layers, 32 μm thick on anterior portion and 48 μm thick in tail region; head bearing six papillae; pharynx represented by a cuticular tube, 17 to 26 μm in diameter, surrounded by pharyngeal tissue reaching 130 μm in diameter; pharyngeal tube meets a blind sac which is directed anteriorly; no genital organs were observed; cross sections show six muscle fields separated by longitudinal chords.

Comments — Unfortunately, the type specimen of *A. restiformis* could not be located in the Armed Forces Institute of Pathology Medical Museum. Apparently the specimen was either lost or misplaced as a result of many physical moves of the museum.[365a] Thus, any interpretations on the identity of this nematode have to be based on the published accounts of Leidy[364] and Stiles.[365] As the latter author mentioned, the presence of six longitudinal chords, cross fibers in the cuticle, cuticular lining of the pharynx, and a suggestion of a blind intestinal sac are characters found in members of the Mermithidae. However, the amount of tissue surrounding the pharyngeal tube and the association of this tissue with the intestine suggests that the specimen might belong to another nematode group. Also, the large size (66cm) would certainly be a record for the mermithids, although it lies within the realm of possibility. The only known species that comes near to attaining this size in the U.S. is *Mermis nigrescens* (Duj.) (syn. *subnigrescens* Cobb) which reaches 16.3 cm[125] and possesses a pair of lip papillae clearly lacking in the species under question (compare Figures 85 and 92). Leidy[364] also stated that the pharynx opened into a straight cylindrical intestine, a character not mentioned in the redescription of Stiles.[365] However, in a juvenile mermithid of this size, the pharynx (= stichosome) and intestine are separate and not connected with one another.

Assuming that the information given by the patient is correct (an assumption questioned by Stiles[365] who stated, ''Further, the idea that an error has occurred in interpreting this worm as a parasite of man seems to gain support, for it would be exceedingly difficult to explain the presence of a Mermithidae in the bladder''), there is the question of the parasite's identity. Although Stiles[365] considered this specimen a mermithid, Leidy[364] did not, and neither description gives conclusive evidence that the worm belongs to this family. Could it possibly have been a juvenile of the giant kidney worm *Diocytophyma renale* (Goeze)? This latter nematode occurs in the kidney and sometimes other organs of both wild and domestic animals throughout the world, and at least ten cases have been reported from humans. The parasites are reddish in color, and the males range from 14 to 45 cm in length and females from 20 to 100 cm.[366] The cuticle is striated, and the pharynx is long and narrow. The worms may attempt to escape down the ureter and produce acute uremic poisoning or may succeed in escaping from the urethra, causing the urine to contain blood and pus.[367] These symptoms are almost identical to the following that were described from the patient who passed *A. restiformis*. ''Previous to its passage, his urine was of a milky hue and some time subsequently of a yellow cast and slightly tinged with blood and mingled with mucus.''[364] Thus, it seems possible that *A. restiformis* may have been a juvenile of the giant kidney worm.

C. Case 3: Mermithid from a Women in Australia[368]

Location — ''. . . said to have been passed by a woman thought to be suffering from uterine cancer, whether per vaginam, rectum or urethrum, not stated.''[368]

Description — [368]Pinkish flesh color, about 56 cm long; greatest width about 1mm. The present author examined the specimen, agrees that it is an immature mermithid and adds the following supplementary description (Figures 83 and 84): lip papillae absent; six head papillae; cross fibers in the cuticle; pharyngeal tube present, relatively wide as compared to other mermithids; amphids not distinct; a minute spike on the tip of the tail; protein platelets present in the pseudocoelom (from 41 to 77 μm in diameter) and similar in shape and form to those observed in *Hydromermis* sp.[358]

Comments — Unfortunately, the data surrounding its occurrence are rather meager, and definite conclusions as to its parasitic role in humans cannot be made.

D. Case 4: Mermithid from the Urethra of an Alcoholic Individual

Location — Neveu-Lemaire[369] mentioned a case where a mermithid (designated as

A. restiformis) was found in the urethra of an alcoholic individual, similar to the case cited previously. Neveu-Lemaire considered it to be a case of accidental parasitism; however, further details and a description are lacking.

E. Case 5: Mermithid Said to Occur in the Intestine of a Child

Location — A nematode identified as *Agromermis hominis* by Leon[370] was collected from the feces of a 5-year-old girl from Ibarra, Ecuador. She had just been given a treatment for *Ascaris.*

Description — Body whitish, 210 mm long; greatest width, 0.8 mm; width 1 mm from the head, 0.15 mm; width 1 mm from the tail end, 0.5 mm; pharyngeal tube 0.8 mm long and 0.011 mm in diameter; the trophosome extends to 0.300 mm from the head end; cuticle with fine cross fibers.

Comments — Leon's description fits that of a mermithid nematode. Leon considered the worm as an accidental parasite in the intestine of the child and suggested that it was ingested with food or water. It is not known if postparasitic juvenile mermithids can survive the conditions found in the digestive tract of warm-blooded vertebrates; however, if the data are correct, they would suggest that such survival is possible. Whether it was parasitic or simply maintained itself in this habitat remains unanswered.

F. Case 6: Mermithid from a Girl in the Belgium Congo[371]

Location — Found in the urethra of a 10-year-old girl who had suffered violent, hypogastric pain during the preceding night. Her urine was still slightly bloody 3 days later.

Description — Body yellow in color, 25 cm long, 80 μm in the widest portion; head rounded, mouth terminal; cuticle 40 μm thick in the widest portion, with superficial cross fibers; pharyngeal tube 6.4 mm long; six cephalic papillae; amphids not visible; trophosome extends to about 800 μm from the anterior end and occupies most of the body cavity. The tail is rounded, and anus and genital organs are lacking.

Comments — The above description fits that of a mermithid nematode, and there is no doubt that the authors have correctly placed it in the genus *Agamomermis*. The authors point out that records show that children and disturbed adults will often voluntarily place objects in their urethra or genital openings, and this case is explained as happening under similar circumstances.

G. Case 7: Mermithid Recovered from Urine

Location — The specimen, which the author received from Beltsville, Maryland, was simply labeled "from urine" and evidently was sent to B. G. Chitwood after being collected by Dr. Hoffman in Puerto Rico on August 5, 1932. It had been labeled "*Hexamermis hoffmani* n. sp." on the slides, but since a published description never appeared, the name is invalid.

Description (Figure 86) — Length 56 mm; greatest width, 0.336 mm; no lip papillae; six head papillae; no cross fibers in the cuticle; minute scar on the tip of the tail. The specimen is a postparasitic juvenile mermithid surrounded by a loose cuticle. A small rudimentary genital opening indicated that the specimen was either an immature male or intersex.

Comments — Again, the brevity of information concerning the location of the worm makes it impossible to state whether it had spent a period of time in the human body.

H. Case 8:

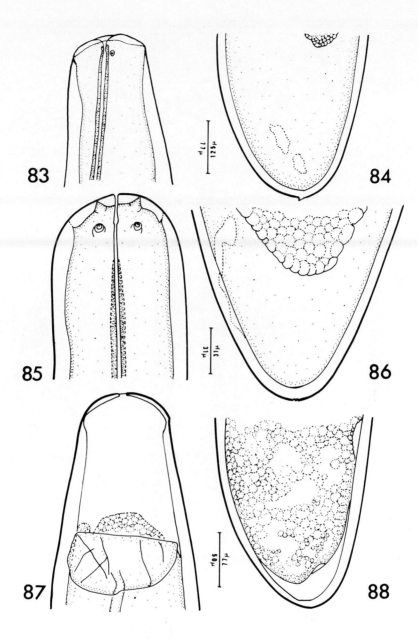

FIGURE 83 to 88.   (83) Head of a mermithid supposedly passed by a woman in Australia. (84) Tail of a mermithid said to have been passed by a woman in Australia. (85) Head of a postparasitic juvenile female *Mermis nigrescens* reared from an insect. (86) Tail of a mermithid collected from "urine" in Puerto Rico. (87) Head of a mermithid supposed passed in the feces of a child in Ecuador. (88) Tail of a mermithid said to have been passed in the feces of a child in Ecuador.

H. Case 8: Mermithid Reportedly Passed from the Intestine of a Child

Location — From the feces of a boy in Ibarra, Ecuador; collected on May 26, 1942.

Description (Figures 87 and 88) — A juvenile mermithid; length 210 mm; greatest width, 0.507 mm; width at head 1 mm from tip, 0.288 mm; width at tail 1 mm from tip, 0.480 mm; pharynx extends over 6 mm from the head end, and the diameter of the pharyngeal tube is 0.016 mm; tip of trophosome to head end, 0.516 mm; no lip papillae; six head papillae; cuticle with faint cross striations; no tail spike; trophosome extends almost to tip of tail.

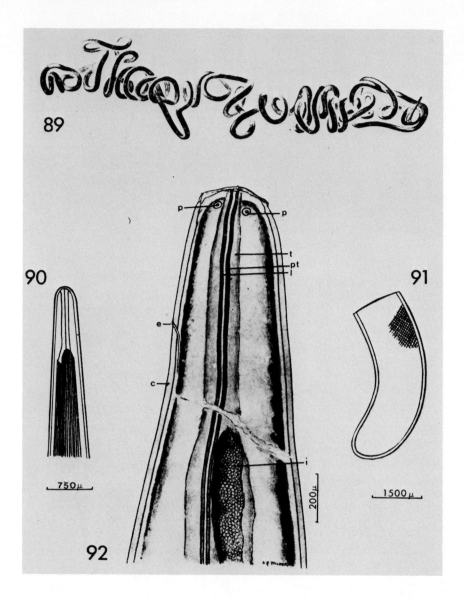

FIGURE 89 to 92.   (89) A copy of one of Stiles' figures[365] (slightly modified) from his study on the redescription of *Agamomermis restiformis* Leidy. (89) Entire specimen of *Agamomermis restiformis* (approximately natural size). (90) Leidy's drawing of the head of *Agamomermis restiformis* copied by Stiles. (91) Leidy's drawing of the tail of *Agamomermis restiformis* copied by Stiles. (92) Anterior end of *Agamomermis restiformis*. c, cuticle; e, excretory pore (?); i, intestine; l, lumen of pharynx; p, papilla; pt, pharyngeal tube; t, tissue surrounding pharyngeal tube.

Comments — This mermithid was obtained from Dr. Leon of Quito, Ecuador. Although the conditions under which it was found were very similar to the published account of *A. hominis*,[370] it is clearly different and was collected 2 years earlier than the case mentioned above.

In most cases described here, the data pertaining to the occurrence of the parasite are such that it cannot be ascertained if human parasitism actually occurred. In Cases 5 and 8, for instance, the worms were obtained from feces and could simply have been ingested and passed unharmed through the digestive tract of the patient or they could have emerged from parasitized sarcophagus insects that had visited the feces. The latter explanation could also account for the presence of mermithids in urine (Case 7 and possibly Cases 4 and 6 if worms were found in the urine and assumed to have been passed from the urethra). A parasitized insect could have fallen into the toilet bowl or free-living mermithids could have already been in the area. A case illustrating the latter possibility was mentioned by Prudhoe,[371a] regarding a small boy who was said to have passed two mermithids (later identified as adult females of *Mermis nigrescens*). Apparently the boy had urinated among some bushes, and the nematodes were found in the area he had covered. The boy's mother concluded he had passed them while urinating. This occurred in June when the females of *Mermis nigrescens* are found near the surface of the soil.

Only in Case 2, where the description of the location of the parasite in the patient was rather extensive and detailed, would there be less chance of an error in observation. However, in this case there was a question regarding the identity of the parasite which cannot be resolved until the specimen is located and reexamined. Even then, it may be in such a condition that a complete identification is impossible.

Needless to say, the question of accidental human infection by mermithid nematodes cannot be definitely answered at this time and should only be accepted as fact when proven experimentally or when parasites are found developing *in situ* in the human organism.

Chapter 9

FUTURE PROSPECTS

Of the several dozen entomogenous nematodes discussed in this book, a few have already been used as biological control agents, and others will undoubtedly follow them. Just how many potentially useful parasites still await discovery is difficult to say, but the number is probably not small.

How well we utilize these nematodes to our advantage depends on our ability to recognize a good candidate when one is discovered. After initial discovery, several steps logically follow. First, some technique must be developed to culture the nematode, at least long enough to obtain the infective stages for experimental infections. Then the life history and effect on host can be investigated. If the candidate still looks promising, then a more permanent method of culture and of storing the parasite can be developed. Eventually, host range studies, survival rate in nature, and behavior of the infective stages can be examined.

Finally, the nematodes can be applied against a particular insect or group of insects, preferably starting in the greenhouse or on small field plots. Afterwards, release can be made on a larger scale. Success will largely depend on a correct analysis of the environment and determining, on the basis of laboratory studies, just what effect that environment will have on the parasite.

Of course, it is impossible to take into consideration all chemical, physical, and biological factors that might adversely affect a nematode parasite since we still do not know what these factors comprise, and therefore luck and imagination are still important aspects in any program. The process of obtaining official approval for the mass release of nematodes in the environment is as important a factor as developing a method of mass rearing the parasite.

If the parasite gives a high degree of pest control and can be economically reared, then a commercial company may be interested in marketing the product. Such is the case with the mermithid *Romanomermis culicivorax* and was the case with *Neoaplectana carpocapsae*. However, other agencies can be as effective in producing the parasites for field testing, as has been clearly demonstrated on several instances.

Any entomogenous nematode that kills or sterilizes one or more pest insects and can be easily grown in the laboratory has potential as a biological control agent. Good candidates like the neoaplectanids, heterorhabditids, and neotylenchids have the advantage of being able to be cultured in vitro. However, this is not always a problem, as is demonstrated by the in vivo rearing methods for mermithid nematodes.

This is a new and exciting field of study which is rapidly developing and certainly contains more possibilities than ever could be covered here.

REFERENCES

1. Poinar, G. O., Jr., *Entomogenous Nematodes,* E. J. Brill, Leiden, 1975, 317.
2. Poinar, G. O., Jr., *CIH Key to the Groups and Genera of Nematode Parasites of Invertebrates,* Commonwealth Agricultural Bureaux, Farnham Royal, England, 43, 1977.
3. Nickle, W. R., A contribution to our knowledge of the Mermithidae (Nematoda), *J. Nematol.,* 4, 113, 1972.
4. Tsai, Y. and Grundmann, A. W., *Reesimermis nielseni* gen. et sp. n. (Nematoda: Mermithidae) parasitizing mosquitoes in Wyoming, *Proc. Helminthol. Soc. Wash.,* 36, 61, 1969.
5. Ross, J. F. and Smith, S. M., A review of the mermithid parasites (Nematoda: Mermithidae) described from North American mosquitoes (Diptera: Culicidae) with descriptions of three new species, *Can. J. Zool.,* 54, 1084, 1976.
6. Petersen, J. J., Comparative biology of the Wyoming and Louisiana populations of *Reesimermis nielseni,* parasitic nematodes in mosquitoes, *J. Nematol.,* 8, 273, 1976.
7. Poinar, G. O., Jr. and Hess, R., Structure of the pre-parasitic juveniles of *Filipjevimermis leipsandra* and some other Mermithidae (Nematoda), *Nematologica,* 20, 163, 1974.
8. Gordon, R., Bailey, C. H., and Barber, J. M., Parasitic development of the mermithid nematode *Reesimermis nielseni* in the larval mosquito *Aedes aegypti, Can. J. Zool.,* 52, 1923, 1974.
9. Poinar, G. O., Jr. and Otieno, W. A., Evidence of four molts in the Mermithidae, *Nematologica,* 20, 370, 1974.
10. Petersen, J. J., Chapman, H. C., and Woodard, D. B., Bionomics of a mermithid nematode of larval mosquitoes in southwestern Louisiana, *Mosq. News,* 28, 346, 1968.
11. Petersen, J. J., Development and fecundity of *Reesimermis nielseni,* a nematode parasite of mosquitoes, *J. Nematol.,* 7, 211, 1975.
12. Otieno, W. A., Studies on the Parasitic Development of *Reesimermis nielseni* Tsai and Grundmann (Mermithidae) in *Culex pipiens* (Culicidae), M.S. thesis, University of California, Berkeley, 1975, 81.
13. Poinar, G. O., Jr., and Hess, R., *Romanomermis culicivorax:* morphological evidence of transcuticular uptake, *Exp. Parasitol.,* 42, 27, 1977.
14. Petersen, J. J. and Willis, O. R., A two year survey to determine the incidence of a mermithid nematode in mosquitoes in Louisiana, *Mosq. News,* 31, 558, 1971.
15. Petersen, J. J. and Willis, O. R., Some factors affecting parasitism by mermithid nematodes in southern house mosquito larvae, *J. Econ. Entomol.,* 63, 175, 1970.
16. Brown, B. J. and Platzer, E. G., The effects of temperature on the infectivity of *Romanomermis culicivorax, J. Nematol.,* 9, 166, 1977.
17. Hughes, D. S. and Platzer, E. G., Temperature effects on the parasitic phase of *Romanomermis culicivorax* in *Culex pipiens, J. Nematol.,* 9, 173, 1977.
18. Levy, R. and Miller, T. W., Jr., Thermal tolerance of *Romanomermis culicivorax,* a nematode parasite of mosquitoes, *J. Nematol.,* 9, 259, 1977.
19. Galloway, T. D. and Brust, R. A., Effects of temperature and photoperiod on the infection of two mosquito species by the mermithid, *Romanomermis culicivorax, J. Nematol.,* 9, 218, 1977.
20. Nickle, W. R., Toward the commercialization of a mosquito mermithid, in Proc. 1st Int. Colloq. Invertebrate Pathology, Kingston, Canada, 1976, 241.
21. Petersen, J. J., Factors affecting sex ratios of a mermithid parasite of mosquitoes, *J. Nematol.,* 4, 83, 1972.
22. Chen, P.-S., A study on *Reesimermis nielseni* for control of *Culex pipiens fatigans* in Taiwan, *Bull. Inst. Zool. Acad. Sin.,* 15, 21, 1976.
23. Mitchell, C. J., Chen, P.-S., and Chapman, H. C., Exploratory trials utilizing a mermithid nematode as a control agent for *Culex* mosquitoes in Taiwan, *J. Formosan Med. Assoc.,* 73, 241, 1974.
24. Levy, R. and Miller, T. W., Jr., Susceptibility of the mosquito nematode, *Romanomermis culicivorax* (Mermithidae) to pesticides and growth regulators, *Environ. Entomol.,* 6, 447, 1978.
25. Levy, R. and Miller, T. W., Jr., Personal correspondence, 1978.
26. Platzer, E. G. and Brown, B. J., Physiological ecology of *Reesimermis nielseni,* in Proc. 1st Int. Colloq. Invertebrate Pathology, Kingston, Canada, 1976, 263.
27. Brown, B. J. and Platzer, E. G., Salts and the infectivity of *Romanomermis culicivorax, J. Nematol.,* 10, 53, 1978.
28. Savage, K. E. and Petersen, J. J., Observations of mermithid nematodes in Florida mosquitoes, *Mosq. News,* 31, 218, 1971.
29. Hansen, E. L. and Hansen, J. W., Parasitism of *Simulium damnosum* by *Romanomermis culicivorax, I.R.C.S. Med. Sci.,* 4, 508, 1976.
30. Kerdpibule, V., Deesin, T., Sucharit, S., and Harinasuta, C., A preliminary study on the control of *Mansonia uniformis* by nematode parasitism *(Reesimermis nielseni). Southeast Asian J. Trop. Med. Public Health,* 5, 150, 1974.

31. **Hansen, E. L. and Hansen, J. W.**, personal communication, 1977.
32. **Petersen, J. J. and Willis, O. R.**, Experimental release of a mermithid nematode to control floodwater mosquitoes in Louisiana, *Mosq. News,* 36, 339, 1976.
33. **Galloway, T.**, Application of a mermithid nematode *(Reesimermis nielseni* Tsai and Grundmann) from Louisiana for mosquito control in Manitoba, *Proc. Alberta Mosq. Abat. Symp.,* 191, 1975.
34. **Chapman, H. C., Clark, T. B., and Petersen, J. J.**, Protozoa, nematodes and viruses of anophelines, *Misc. Pub. Entomol. Soc. Am.,* 7, 134, 1970.
35. **Brown, B. J., Platzer, E. G., and Hughes, D. S.**, Field trials with the mermithid nematode, *Romanomermis culicivorax,* in California, *Mosq. News,* 37, 603, 1977.
36. **Otieno, W. A.**, The Pathology and Host-Parasite Relationship of the Entomogenous Nematode *Romanomermis culicivorax* Ross and Smith (Mermithidae) in a Mosquito Host *Culex pipiens* Say (Culicidae): Its Application in Integrated Control, Ph.D. thesis, University of California, Berkley, 1977, 241.
37. **Kurihara, T.**, Population behaviour of *Reesimermis nielseni,* a nematode parasite of mosquitoes, with notes on the attraction of infective stage nematodes by mosquito larvae, *Culex pipiens molestus, Jpn. J. Parasitol.,* 25, 8, 1976.
38. **Ignoffo, C. M. et al.**, Susceptibility of aquatic vertebrates and invertebrates to the infective stage of the mosquito nematode *Reesimermis nielseni, Mosq. News,* 33, 599, 1973.
39. **Petersen, J. J., Chapman, H. C., and Willis, O. R.**, Fifteen species of mosquitoes as potential hosts of a mermithid nematode *Romanomermis* sp., *Mosq. News,* 29, 198, 1969.
40. **Finney, J. R.**, The penetration of three simuliid species by the nematode *Reesimermis nielseni, Bull. WHO,* 52, 235, 1975.
41. **Petersen, J. J. and Willis, O. R.**, Procedures for the mass rearing of a mermithid parasite of mosquitoes, *Mosq. News,* 32, 226, 1972.
42. **Petersen, J. J.**, Factors affecting the mass rearing of *Reesimermis nielseni,* a nematode parasite of mosquitoes, *J. Med. Entomol.,* 10, 75, 1973.
43. **Sanders, R. D., Stokstad, E. L. R., and Malatesta, C.**, Axenic growth of *Reesimermis nielseni* (Nematoda: Mermithidae) in insect tissue culture media, *Nematologica,* 19, 567, 1973.
44. **Roberts, D. W. and van Leuken, W.**, Limited growth of a mermithid nematode in association with tissue culture cells, paper presented at 1st Int. Colloq. Insect Pathology Microbial Control, Oxford, England, 1973, 91.
45. **Finney, J. R.**, The development of *Romanomermis culicivorax* in in vitro culture, *Nematologica,* 23, 479, 1978.
46. **Petersen, J. J. and Willis, O. R.**, Results of preliminary field applications of *Reesimermis nielseni* (Mermithidae: Nematoda) to control mosquito larvae, *Mosq. News,* 32, 312, 1972.
47. **Petersen, J. J., Hoy, J. B., and O'Berg, A. G.**, Field application of *Reesimermis nielseni* (Mermithidae: Nematoda) against mosquito larvae in California rice fields, *Calif. Vector Views,* 19, 47, 1972.
48. **Petersen, J. J., Steelman, C. D., and Willis, O. R.**, Field parasitism of two species of Louisiana rice field mosquitoes by a mermithid nematode, *Mosq. News,* 33, 573, 1973.
49. **Petersen, J. J.**, Status and future of mermithid nematodes as control agents of mosquitoes, in Proc. 1st Int. Colloq. Invertebrate Pathology, Kingston, Ontario, 1976, 236.
50. **Petersen, J. J.**, Penetration and development of the mermithid nematode *Reesimermis nielseni* in eighteen species of mosquitoes, *J. Nematol.,* 7, 211, 1975.
51. **Chapman, H. C.**, Biological control of mosquito larvae, *Annu. Rev. Entomol.,* 19, 33, 1974.
52. **Petersen, J. J. and Willis, O. R.**, Experimental release of a mermithid nematode to control *Anopheles* mosquitoes in Louisiana, *Mosq. News,* 34, 316, 1974.
53. **Petersen, J. J. and Willis, O. R.**, Establishment and recycling of a mermithid nematode for the control of mosquito larvae, *Mosq. News,* 35, 526, 1975.
54. **Levy, R. and Miller, T. W., Jr.**, Experimental release of a mermithid to control mosquitoes breeding in sewage settling tanks, *Mosq. News,* 37, 410, 1977.
55. **Levy, R. and Miller, T. W., Jr.**, Experimental release of *Romanomermis culicivorax* (Mermithidae: Nematoda) to control mosquitoes breeding in southwest Florida, *Mosq. News,* 37, 483, 1977.
56. **Levy, R., Hertlein, B. C., Petersen, J. J., and Miller, T. W., Jr.**, Aerial application of *Romanomermis culicivorax* (Mermithidae: Nematoda) to control *Anopheles* and *Culex* mosquitoes in Southwest Florida, *Mosq. News* in press.
57. **Levy, R., Murphy, L. J., Jr., and Miller, T. W., Jr.**, Effects of a simulated aerial spray system on a mermithid parasite of mosquitoes, *Mosq. News,* 36, 498, 1976.
58. **Levy, R., Cornell, J. A., and Miller, T. W., Jr.**, Application of a mermithid parasite with an aerial spray system, *Mosq. News,* 37, 512, 1977.
59. **Petersen, J. J.**, personal correspondence, 1978.
60. **Obiamiwe, B. A.**, The life cycle of *Romanomermis* sp. (Nematoda: Mermithidae) a parasite of mosquitoes, *Trans. R. Soc. Trop. Med. Hyg.,* 63, 18, 1969.

61. Obiamiwe, B. A. and MacDonald, W. W., A new parasite of mosquitoes *Reesimermis muspratti* sp. nov. (Nematoda: Mermithidae), with notes on its life cycle, *Ann. Trop. Med. Parasitol.*, 67, 439, 1973.

62. Petersen, J. J., Biology of *Octomyomermis muspratti,* a parasite of mosquitoes as it relates to mass production, *J. Invertebr. Pathol.*, 30, 155, 1977.

63. Petersen, J. J., Effects of host size and parasite burden on sex ratio in the mosquito parasite *Octomyomermis muspratti, J. Nematol.*, 9, 343, 1977.

64. Petersen, J. J., Effects of male-female ratios on mating and egg production in *Octomyomermis muspratti* (Mermithidae: Nematoda), *J. Invertebr. Pathol.*, 31, 103, 1978.

65. Muspratt, J., Observations on the larvae of treehole breeding culicini (Diptera: Culicidae) and two of their parasites, *J. Entomol. Soc. South Afr.*, 8, 13, 1945.

66. Muspratt, J., Technique for infecting larvae of *Culex pipiens* complex with a mermithid nematode for culturing the latter in the laboratory, *Bull. WHO*, 33, 140, 1965.

67. Laird, M., Microbial control of arthropods of medical importance, in *Microbial Control of Insects and Mites,* Burges, H. D., Ed., Academic Press, New York, 1971, 134.

68. Poinar, G. O., Jr., and Sanders, R. D., Description and biology of *Octomyomermis troglodytis* sp. n. (Nematoda: Mermithidae) parasitizing the Western treehole mosquito, *Aedes sierrensis* (Ludlow) (Diptera: Culicidae), *Proc. Helminthol. Soc. Wash.*, 41, 37, 1974.

69. Welch, H. E., *Hydromermis churchillensis* n. sp. (Nematoda: Mermithidae) a parasite of *Aedes communis* (De G.) from Churchill, Manitoba, *Can. J. Zool.*, 38, 465, 1960.

70. Jenkins, D. W. and West, A. S., Mermithid nematode parasites in mosquitoes, *Mosq. News*, 14, 138, 1954.

71. Welch, H. E., Notes on the identities of mermithid parasites of North American mosquitoes and a redescription of *Agamomermis culicis* Stiles, 1903, *Proc. Helminthol. Soc. Wash.*, 27, 203, 1970.

72. Petersen, J. J., Chapman, H. C., and Woodard, D. B., Preliminary observations on the incidence and biology of a mermithid nematode of *Aedes sollicitans* (Walker) in Louisiana, *Mosq. News*, 27, 493, 1967.

73. Petersen, J. J., and Willis, O. R., Incidence of *Agamomermis culicis* Stiles (Nematoda: Mermithidae) in *Aedes sollicitans* in Louisiana during 1967, *Mosq. News*, 29, 87, 1969.

74. Stiles, C. W., A parasitic roundworm (*Agamomermis culicis* n. g., n. sp.) in American mosquitoes *(Culex sollicitans), Hygiene Lab. U.S. Public Health, Marine Hosp. Serv. Bull.*, 13, 15, 1903.

75. Petersen, J. J. and Chapman, H. C., Parasitism of *Anopheles* mosquitoes by a *Gastromermis* sp. (Nematoda: Mermithidae) in southwestern Louisiana, *Mosq. News*, 30, 420, 1970.

76. Woodard, D. B. and Fukuda, T., Laboratory resistance of the mosquito, *Anopheles quadrimaculatus* to the mermithid nematode *Diximermis peterseni, Mosq. News*, 37, 192, 1977.

77. Poinar, G. O., Jr., *Hydromermis conopophaga* n. sp., parasitizing midges (Chironomidae) in California, *Ann. Entomol. Soc. Am.*, 61, 593, 1968.

78. Hominick, W. M. and Welch, H. E., Morphological variation in three mermithids (Nematoda) from Chironomidae (Diptera) and a reassessment of the genera *Gastromermis* and *Hydromermis, Can J. Zool.*, 49, 807, 1971.

79. Hominick, W. M. and Welch, H. E., Synchronization of life cycles of three mermithids (Nematoda) with their chironomid (Diptera) hosts and some observations on the pathology of the infections, *Can. J. Zool*, 49, 975, 1971.

80. Chapman, J. and Ecke, D. H., Study of a population of chironomid midges (*Tanytarsus*) parasitized by mermithid nematodes in Santa Clara County, California, *Calif. Vector Views*, 16, 83, 1969.

81. Hagmeier, A., Beiträge zur Kenntnis der Mermithiden. I. Biologische Notizen und systematische Beschreibung einiger alter und neuer Arten, *Zool. Jahrb. Abt. Syst. Oekol. Geogr. Tiere*, 32, 521, 1912.

82. Götz, P., Der Einfluss Unterschiedlicher Befallsbedingungen auf die Mermithogene Intersexualität von *Chironomus* (Dipt.), *Z. Parasitenkd.*, 24, 484, 1964.

83. Wülker, W., Der Mechanismus des Eindringens parasitarer Mermithiden (Nematoda) in *Chironomus* — Larven (Dipt., Chironomidae), *Z. Parasitenkd.*, 26, 29, 1965.

84. Wülker, W., Untersuchungen über die Intersexualität der Chironomiden (Dipt.) nach *Paramermis* — Infektion, *Arch. Hydrobiol. Suppl.*, 25, 127, 1961.

85. Wülker, W., Prospects for biological control of pest Chironomidae in the Sudan, *WHO/EBL*, 23, 1963.

86. Wülker, W., *Gastromermis rosea* (Nematodes). Embryonalentwicklung, *Inst. Wissensch. Cinemat.*, E1563/1970, 3, 1971.

87. Wülker, W. Parasitismus des Nematoden *Gastromermis rosea* in *Chironomus anthracinus* (Diptera), *Inst. Wissensch. Cinemat.*, C1024/1970, 3, 1974.

88. Johnson, A. A., Life History Studies on *Hydromermis contorta* (Kohn) a Nematode Parasite of *Chironomus plumosus* (L.), Ph.D. thesis, University of Illinois, Urbana, 1955, 92.

89. Kohn, F. G., Einiges über *Paramermis contorta* (v. Linstow) = *Mermis contorta* v. Linstow, *Arb. Zool. Inst. Univ. Wien*, 15, 213, 1905.

90. Poinar, G. O., Jr. and Tourenq, J. N., On the occurrence of *Hydromermis contorta* (Kohn) (Nematodea) parasitizing midges (Chironomidae) in the Camargue, *Ann. Limnol.*, 8, 41, 1972.

91. McCauley, V. J. E., Mermithid (Nematoda) parasites of Chironomidae (Diptera) in Marion Lake, British Columbia, *J. Invertebr. Pathol.*, 22, 454, 1973.

92. Phelps, R. J. and DeFoliart, G. R., Nematode parasitism of Simuliidae, *Wis. Agric. Exp. Stn. Res. Bull.*, 245, 78, 1964.

93. Rubtsov, I. A., Mermithids parasitic in simuliids, *Zool. Zh.*, 42, 1768, 1963.

94. von Daday, E., Adatok a Mermithidae-Család édes vízben élő fajainak ismeretchez, *Math. Termés zettud. Ertesitö Magyar Tudomán. Akad. Budapest*, 29, 450, 1911.

95. Chitwood, B. G., Nomenclatorial notes, *Proc. Helminthol. Soc. Wash.*, 1, 51, 1935.

96. Welch, H. E., New species of *Gastromermis*, *Isomermis* and *Mesomermis* (Nematoda: Mermithidae) from black fly larvae, *Ann. Entomol. Soc. Am.*, 55, 535, 1962.

97. Ebsary, B. A. and Bennett, G. F., Redescription of *Neomesomermis flumenalis* (Nematoda) from blackflies in Newfoundland, *Can. J. Zool.*, 52, 65, 1974.

98. Ebsary, B. A. and Bennett, G. F., Molting and oviposition of *Neomesomermis flumenalis* (Welch, 1962) Nickle, 1972, a mermithid parasite of black flies, *Can. J. Zool.*, 51, 637, 1973.

99. Mokry, J. E. and Finney, J. R., Mermithids from Adult Blackflies, Ann. Rep. Research Unit on Vector Pathology, Memorial University of Newfoundland, St. John's, 1976, 14.

100. Anon., 3rd Biannual Rep. Research Unit on Vector Pathology, Memorial University of Newfoundland, St. John's, 1973,

101. Molloy, D. and Jamnback, H., Laboratory transmission of mermithids parasitic in black flies, *Mosq. News.*, 35, 337, 1975.

102. Ezenwa, A. O. and Carter, N. E., Influence of multiple infections on sex ratios of mermithid parasites of blackflies, *Environ. Entomol.*, 4, 142, 1975.

103. Condon, W. J. and Gordon, R., Some effects of mermithid parasitism on the larval blackflies *Prosimulium mixtum fuscum* and *Simulium venustum*, *J. Invertebr. Pathol.*, 29, 56, 1977.

104. Ebsary, B. A. and Bennett, G. F., The occurrence of some endoparasites of blackflies (Diptera: Simuliidae) in insular Newfoundland, *Can. J. Zool.*, 53, 1058, 1975.

105. Ebsary, B. A. and Bennett, G. V., Studies on the bionomics of mermithid nematode parasites of blackflies in Newfoundland, *Can. J. Zool.*, 53, 1324, 1975.

106. Ezenwa, A. O., Ecology of Simuliidae, Mermithidae, and Microsporida in Newfoundland freshwaters, *Can. J. Zool.*, 52, 557, 1974.

107. Ezenwa, A. O., Studies on host-parasite relationships of Simuliidae with mermithids and microsporidans, *J. Parasitol.*, 60, 809, 1974.

108. Bailey, C. H. and Gordon, R., Observations on the occurrence and collection of mermithid nematodes from blackflies (Diptera: Simuliidae), *Can. J. Zool.*, 55, 148, 1977.

109. Ezenwa, A. O., Mermithid and microsporidan parasitism of black flies (Diptera: Simuliidae) in the vicinity of Churchill Falls, Labrador, *Can. J. Zool.*, 51, 1109, 1973.

110. Lewis, D. J. and Bennett, G. F., The blackflies (Diptera: Simuliidae) of insular Newfoundland. III. Factors affecting the distribution and migration of larval simuliids in small streams in the Avalon Peninsula, *Can. J. Zool.*, 53, 114, 1975.

111. Bailey, C. H., Gordon, R., and Mills, C., Laboratory culture of the free living stages of *Neomesomermis flumenalis*, a mermithid nematode parasite of Newfoundland blackflies (Diptera: Simuliidae), *Can. J. Zool.*, 55, 391, 1977.

112. Molloy, D. and Jamnback, H., A larval black fly control field trial using mermithid parasites and its cost implications, *Mosq. News*, 37, 104, 1977.

113. Mondet, B., Poinar, G. O., Jr., and Bernadou, J., Etude du parasitisme des simulies (Diptera, Simuliidae) par des Mermithidae (Nematoda) en Afrique de L'Ouest. IV. Description de *Isomermermis lairdi* n. sp., parasite de *Simulium damnosum*, *Can. J. Zool.*, 55, 2011, 1977.

114. Mondet, B., Pendriez, B., and Bernadou, J., Etude du parasitisme des simulies (Diptera) par des Mermithidae (Nematoda) en Afrique de l'Ouest. I. Observations preliminaires sur un cours d'eau temporaire de savane, *Cah. ORSTOM Ser. Entomol. Med. Parasitol.*, 14, 141, 1976.

115. Mondet, B., Berl, D., and Bernadou, J., Etude du parasitisme des simulies (Diptera: Simuliidae) par des Mermithidae (Neimatoda) en Afrique de l'Ouest. III. Elevage de *Isomermis* sp. et infestation en laboratoire de *Simulium damnosum*. *Cah. ORSTOM Ser. Entomol. Med. Parasitol.*, in press.

115a. Walsh, personal correspondence.

115b. Hansen, E. L. and Hansen, J. W., personal correspondence.

116. Poinar, G. O., Jr., Remillet, M., and van Waerebeke, D., *Mermis changodudus* sp. n. (Mermithidae) a nematode parasite of *Heteronychus* beetles (Scarabaeidae) in Madagascar, *Nematologica*, 24, 100, 1978.

117. **Cobb, N. A., Steiner, G., and Christie, J. R.**, *Agamermis decaudata* Cobb, Steiner and Christie: a nema parasite of grasshoppers and other insects, *J. Agric. Res. (Washington, D. C.)*, 23, 921, 1923.

118. **Christie, J. R.**, *Mermis subnigrescens*, a nematode parasite of grasshoppers, *J. Agric. Res. (Washington, D.C.)*, 55, 353, 1937.

119. **Briand, L. J. and Rivard, I.**, Observations sur *Mermis subnigrescens* Cobb (Mermithidae), nematode parasite des criquets au Quebec, *Phytoprotection*, 45, 73, 1964.

120. **Glaser, R. W. and Wilcox, A. M.**, On the occurrence of a *Mermis* epidemic among grasshoppers, *Psyche*, 25, 12, 1918.

121. **Denner, M. W.**, Biology of the Nematode *Mermis subnigrescens* Cobb, Ph.D. thesis, Iowa State University, Ames, 1968, 138.

122. **Mongkolkiti, S. and Hosford, R. M., Jr.**, Biological control of the grasshopper *Hesperotetlix viridis pratensis* by the nematode *Mermis nigrescens*, *J. Nematol.*, 3, 356, 1971.

123. **Uvarov, B. P.**, *Locusts and Grasshoppers. A Handbook for Their Study and Control*, Imperial Bureau of Entomology, London, 1928, 352.

124. **von Linstow, O. F. B.**, Das genus *Mermis*, *Arch. Mikrosk. Anat. Entwicklungsmech.*, 53, 149, 1898.

125. **Cobb, N. A.**, The species of *Mermis*, a group of very remarkable nemas infesting insects, *J. Parasitol.*, 13, 66, 1926.

126. **von Linstow, O. F. B.**, Uber *Mermis nigrescens* Duj., *Arch. Mikro. Anat. Entwicklungsmech.*, 40, 498, 1892.

127. **Assmuss, E. P.**, Verzeichniss einiger Insecten in denen ich Gordisceen Antraf, *Wien. Entomol. Monatschr.*, 2, 171, 1858.

128. **Harris, W. V.**, The Migratory Locust, Department of Agriculture Pamphlet No. 6, Tanganyika, 1932, 1.

129. **Hayes, W. P. and DeCoursey, I. D.**, Observations of grasshoppers parasitism in 1937, *J. Econ. Entomol.*, 31, 519, 1938.

130. **Diesing, K. M.**, *Systema helminthum*, Vol. 2, Berlin, 1851, 588.

131. **von Siebold, C. T. E.**, Ueber die Fadenwürmer der Insekten, *Entomol. Z.*, 11, 329, 1850.

132. **von Siebold, C. T. E.**, Ueber die Fadenwürmer der Insekten, *Entomol. Z.*, 3, 146, 1842.

133. **Baylis, H. A.**, The larval stages of the nematode *Mermis nigrescens*, *Parasitology*, 38, 10, 1947.

134. **Polozhentsev, P. A.**, The mermithid fauna of *Melolontha hippocastani* Fabr., *Tr. Bashk. Nauchno Issled. Vet. Stantsii*, 3, 301, 1941.

135. **von Siebold, C. T. E.**, Zusatz (to Georg Meissner), *Z. Wiss. Zool. Abt. A*, 7, 141, 1855.

136. **Schultz, O.**, Filarien in paläarktischen Lepidopteren, *Illust. Zeit. Entomol.*, 5, 148, 1900.

137. **Baylis, H. A.**, Observations on the nematode *Mermis nigrescens* and related species, *Parasitology*, 36, 122, 1944.

138. **Milum, V. G.**, A larval mermithid, *Mermis subnigrescens* Cobb, as a parasite of the honeybee, *J. Econ. Entomol.*, 31, 460, 1938.

139. **Poinar, G. O., Jr. and Welch, H. E.**, A new nematode, *Filipjevimermis leipsandra* sp. n. (Mermithidae), parasitic in chrysomelid larvae (Coleoptera), *J. Invertebr. Pathol.*, 12, 259, 1968.

140. **Poinar, G. O., Jr.**, Parasitic development of *Filipjevimermis leipsandra* Poinar and Welch (Mermithidae) in *Diabrotica u. undecimpunctata* (Chrysomelidae), *Proc. Helminthol. Soc. Wash.*, 35, 161, 1968.

141. **Cuthbert, J. F. P.**, Bionomics of a mermithid (nematode) parasite of soil-inhabiting larvae of certain chrysomelids (Coleoptera), *J. Invertebr. Pathol.*, 12, 283, 1968.

142. **Poinar, G. O., Jr., and Hess, R.**, Structure of the pre-parasitic juveniles of *Filipjevimermis leipsandra* and some other Mermithidae (Nematodea), *Nematologica*, 20, 163, 1974.

143. **Poinar, G. O., Jr.**, unpublished data, 1968.

144. **Christie, J. R.**, Life history of *Agamermis decaudata*, a nematode parasite of grasshoppers and other insects, *J. Agric. Res. (Washington, D. C.)*, 52, 161, 1936.

145. **Walton, A.**, A revision of the nematodes of the Leidy's collections, *Proc. Acad. Natl. Sci. Philadelphia*, 79, 49, 1927.

146. **Smith, R. W.**, A field population of *Melanoplus sanguinipes* (Fab.) (Orthoptera: Acrididae) and its parasites, *Can. J. Zool.*, 43, 179, 1965.

147. **Cobb, N. A.**, *Tetradonema plicans* nov. gen. et spec., representing a new family, Tetradonematidae as now found parasitic in larvae of the midge-insect *Sciara coprophila* Lintner, *J. Parasitol.*, 5, 176, 1919.

148. **Hungerford, H. B.**, Biological notes on *Tetradonema plicans* Cobb, a nematode parasite of *Sciara coprophila* Lintner, *J. Parasitol.*, 5, 186, 1919.

149. **Ferris, J. M. and Ferris, V. R.**, Observations on *Tetradonema plicans*, an entomoparasitic nematode, with a key to the genera of the family Tetradonematidae (Nematoda: Trichosyringida), *Ann. Entomol. Soc. Am.*, 59, 964, 1966.

150. **Hudson, K. E.**, The form and function of the oesophageal region of the insect parasite *Tetradonema plicans* (Nematoda: Tetradonematidae), *Parasitology*, 63, 137, 1971.

150a. **Poinar, G. O.,** unpublished observations.
151. **Hudson, K. E.,** Regulation of greenhouse sciarid fly populations using *Tetradonema plicans* (Nematoda: Mermithidea), *J. Invertebr. Pathol.,* 23, 85, 1974.
152. **van Waerebeke, D. and Remillet, M.,** Morphologie et biologie de *Heterogonema ovomasculis* n. sp. (Nematoda: Tetradonematidae) parasite de Nitidulidae (Coleoptera), *Nematologica,* 19, 80, 1973.
153. **Merrill, J. H. and Ford, A. L.,** Life history and habits of two new nematodes parasitic on insects, *J. Agric. Res.,* 6, 115, 1916.
154. **Banks, N. and Snyder, T. E.,** A revision of the nearctic termites with notes on biology and geographical distribution, *U. S. Natl. Mus. Bull.,* 108, 228, 1920.
155. **Davis, J. J.,** Contributions to a knowledge of the natural enemies of *Phyllophaga, Ill. Nat. Hist. Surv. Bull.,* 13, 53, 1919.
156. **Winburn, T. F. and Painter, R. H. ,** Insect enemies of the corn earworm (*Heliothis obsoleta* Fabr.), *J. Kans. Entomol. Soc.,* 5, 1, 1932.
157. **Swain, R. B.,** The association of nematodes of the genus *Diplogaster* with White-Fringed beetles, *J. Econ. Entomol.,* 38, 488, 1945.
158. **Poinar, G. O., Jr.,** Diplogasterid nématodes (Diplogasteridae: Rhabditida) and their relationship to insect disease, *J. Invertebr. Pathol.,* 13, 447, 1969.
159. **Maupas, E.,** Essais d'hybridation chez des nématodes, *Bull. Biol. Fr. Belg.,* 52, 466, 1919.
160. **Hirschmann, H.,** Über das Vorkommen zweier Mundhöhlentypen bei *Diplogaster lheritieri* Maupas und *Diplogaster bioformis* n. sp. und die Entstehung dieser hermaphroditischen Art aus *Diplogaster lheritieri, Zool. Jahrb. Abt. Syst. Oekol. Geogr. Tiere,* 80, 132, 1951.
161. **Goodey, T.,** *Soil and Freshwater Nematodes,* rewritten by J. B. Goodey, Methuen & Co., London, 1963, 554.
162. **Weingärtner, I.,** Versuch einer Neuordnung der Gattung *Diplogaster* Schulze 1857 (Nematoda), *Zool. Jahrb. Abt. Syst. Oekol. Geogr. Tiere,* 83, 248, 1955.
163. **Weiser, J.,** *Nemoci Hmyzu,* Academia, Prague, 1966, 554.
164. **Fedorko, A. and Stanuszek, S.,** *Pristionchus uniformis* sp. n. (Nematoda, Rhabditida, Diplogasteridae), a facultative parasite of *Leptinotarsa decemlineata* Say and *Melolontha melolontha* L. in Poland. Morphology and biology, *Acta Parasitol. Pol.,* 19, 95, 1971.
165. **Sandner, H., Seryczynska, H., and Kamionek, M.,** Preliminary microbiological and ultrastructural investigations of bacteria isolated from *Pristionchus uniformis* Fedorko and Stanuszek, *Bull. Acad. Pol. Sci. Cl. 2,* 20, 567, 1972.
166. **Sandner, H. and Stanuszek, S.,** Nematodes-parasites of the Colorado Beetle (*Leptinotarsa decemlineata* Say), in *Proc. 9th Int. Nematode Symp. Warsaw,* paper no. 1967, 1970, 349.
167. **Fedorko, A.,** Nematodes as factors reducing the populations of Colorado beetle, *Leptinotarsa decemlineata* Say, *Acta Phytophathol. Acad. Sci. Hung.,* 6, 175, 1971.
168. **Niklas, O. F.,** Standorteinflüsse und natürliche Feinde als Begrenzungsfaktoren von *Melolontha* — Larvenpopulationen eines Waldgebietes (Forstamt Lorsch, Hessen) (Coleoptera: Scarabaeidae), *Mitt. Biol. Bundesanst. Land Forstwirtsch. Berlin Dahlem,* 101, 60, 1970.
169. **Swain, R. B.,** The association of nematodes of the genus *Diplogaster* with White-Fringed beetles, *J. Econ. Entomol.,* 38, 488, 1945.
170. **Körner, H.,** Die Nematodenfauna des vergehenden Holzes und ihre Beziehungen zu den Insekten, *Zool. Jahrb. Abt. Syst. Oekol. Geogr. Tiere,* 82, 245, 1954.
171. **Griffith, R.,** The mechanism of transmission of the red ring nematode, *J. Agric. Soc. Trinidad Tobago,* 906, 35, 1968.
172. **Mizuta, Y. and Sato, M.,** Some observations on a parasitic nematode from the silkworm, *Sanshi Kenkyu,* 57, 22, 1965.
173. **Surany, P.,** Diseases and biological control in rhinoceros beetles, *South Pac. Comm. Tech. Pap.,* 128, 62, 1960.
174. **Kurian, C.,** Methods of Control of the Coconut Rhinoceros Beetle, *O. rhinoceros* L. Res. Rep. Central Coconut Res. Stn., Kayangulam, India, 1967.
175. **Poinar, G. O., Jr.,** Nematodes as facultative parasites of insects, *Ann. Rev. Entomol.,* 17, 103, 1972.
176. **Blinova, S. L. and Gurando, E. V.,** *Parasitorhabditis fuchsi* sp. n. (Nematoda, Rhabditidae), a parasite of *Blastophagus minor* Hartig., *Vestn. Zool.,* 2, 50, 1974.
177. **Chitwood, B. G. and Chitwood, M. B.,** *An Introduction to Nematology,* Monumental Printing Co., Baltimore, 1937, 372.
178. **Chitwood, B. G. and Chitwood, M. B.,** *An Introduction to Nematology,* (revised), Monumental Printing Co., Baltimore, 1950, 213.
179. **Turco, C. P., Thames, W. H., Jr., and Hopkins, S. H.,** On the taxonomic status and comparative morphology of species of the genus *Neoaplectana* Steiner (Neoaplectanidae: Nematoda), *Proc. Helminthol. Soc. Wash.,* 38, 68, 1971.
180. **Steiner, G.,** *Neoaplectana glaseri,* n. g., n. sp. (Oxyuridae), a new nemic parasite of the Japanese beetle (*Popillia japonica* Newm.) *J. Wash. Acad. Sci.,* 19, 436, 1929.

181. Filipjev, I. N., Miscellanea Nematologica. I. Eine neue Art der Gattung *Neoaplectana* Steiner nebst Bemerkungen über die systematische Stellung der letzteren, *Mag. Parasitol. Inst. Zool. Acad. URSS*, 4, 229, 1934.

182. Skrjabin, K. I., Schikhobalova, N. P., and Mozgovoi, A. A., *Oxyurates and Ascaridates*, Acad. Nauk USSR, Moscow, 1951, 519.

183. Stanuszek, S., *Neoaplectana feltiae* complex (Nematoda: Rhabditoidea, Steinernematidae) its taxonomic position within the genus *Neoaplectana* and intraspecific structure, *Zesz. Probl. Postepow Nauk Roln.*, 154, 331, 1974.

184. Stanuszek, S., *Neoaplectana feltiae pieridarum* N. ecotype (Nematoda: Rhabditoidea, Steinernematidae) — a parasite of *Pieris brassicae* L. and *Mamestra brassicae* L. in Poland. Morphology and biology, *Zesz. Probl. Postepow Nauk Roln.*, 154, 361, 1974.

185. Poinar, G. O., Jr. and Lindhardt, K., The re-isolation of *Neoaplectana bibionis* Bovien (Nematodea) from Danish bibionids (Diptera) and their possible use as biological control agents, *Entomol. Scand.*, 2, 301, 1971.

186. Poinar, G. O., Jr. and Brooks, W. M., Recovery of the entomogenous nematode, *Neoaplectana glaseri* Steiner from a native insect in North Carolina, *IRCS Med. Sci.*, 5, 473, 1977.

187. Poinar, G. O., Jr., Generation polymorphism in *Neoaplectana glaseri* Steiner (Steinernematidae: Nematoda), redescribed from *Strigoderma arboricola* (Fab.) (Scarabaeidae: Coleoptera) in North Carolina, *Nematologica*, 24, 105, 1978.

188. Poinar, G. O., Jr., Description and taxonomic position of the DD-136 nematode (Steinernematidae, Rhabditoidae) and its relationship to *Neoaplectana carpocapsae* Weiser, *Proc. Helminthol. Soc. Wash.*, 34, 199, 1967.

189. Poinar, G. O., Jr. and Veremchuk, G. V., A new strain of entomopathogenic nematode and the geographical distribution of *Neoaplectana carpocapsae* Weiser (Rhabditida, Steinernematidae), *Zool. Zh.*, 49, 966, 1970.

190. Steiner, G., *Aplectana kraussei* n. sp., eine in der Blattwespe *Lyda* sp. parasitierende Nematodenform, nebst Bemerkungen über das Seitenorgan der parasitischen Nematoden, *Zentralbl. Bakteriol. Parasitenkd. Infektionskr. Hyg. Abt. 1 orig.*, 59, 14, 1923.

191. Travassos, L., Sobre o genera *Oxysomatium*, *Bol. Biol.*, 5, 52, 1927.

192. Sobolev, A. A., Clarification of the taxonomic systems of Rhabditates (Superfamilies Rhabditoidea and Aphelenchoidea) parasitizing insects, *Tr. Gel'mintol. 75 let Skrjabin Akad. Nauk SSSR*, 676, 1953.

193. Mracek, Z., *Steinernema kraussei*, a parasite of the body cavity of the sawfly, *Cephalei abietis*, in Czechoslovakia, *J. Invertebr. Pathol.*, 30, 87, 1977.

194. Weiser, J., *Neoaplectana janickii* in an outbreak of the sawfly *Cephaleia abeitis* in Czechoslovakia, *SIP Newsl.*, 8, 8, 1975.

195. Steiner, G., *Neoaplectana glaseri* n. g., n. sp. (Oxyuridae), a new nemic parasite of the Japanese beetle (*Popillia japonica* Newm.), *J. Wash. Acad. Sci.*, 19, 436, 1929.

196. Glaser, R. W., Studies on *Neoaplectana glaseri*, a nematode parasite of the Japanese beetle (*Popillia japonica*), *N. J. Agric.*, 211, 34, 1932.

197. Gardner, T. R. and Parker, L. B., Investigations of the parasites of *Popillia japonica* and related Scarabaeidae in the Far East from 1929 to 1933, inclusive, *U. S. Dep. Agric. Tech. Bull.*, 738, 36, 1940.

198. Clausen, C. P., Jaynes, H. A., and Gardner, T. R., Further investigations of the parasites of *Popillia japonica* in the Far East, *U. S. Dep. Agric. Tech. Bull.*, 366, 58, 1933.

199. Girth, H. B., Recovery of Japanese beetle parasites introduced in May 1940, *Entomol. News*, 65, 97, 1954.

200. Swain, R. B., Nematode parasites of the white-fringed beetles, *J. Econ. Entomol.*, 36, 671, 1943.

200a. Harlan, D., personal correspondence.

201. Swain, R. B., Littig, K. S., Gordon, M. F., and Saul, L. A., Studies of the nematodes and bacterial diseases found associated with the white-fringed beetle, in White Fringed Beetle Investigations, 2nd Q. Rep., Gulfport, Miss., 1943, 43.

202. Swain, R. B., Littig, K. S., Bartlett, F. J., and Gordon, M. F., Studies on nematodes of the genus *Neoaplectana* — factors in the biological control of the white-fringed beetles, in Special Rep. White-Fringed Beetle Investigations, Gulfport, Miss., 1944, 33.

203. Glaser, R. W., McCoy, E. E., and Girth, H. B., The biology and economic importance of a nematode parasitic in insects, *J. Parasitol.*, 26, 479, 1940.

204. Dutky, S. R., Investigation of the Diseases of the Immature Stages of the Japanese Beetle, Ph.D. thesis, Rutgers University, New Brunswick, N.J., 1937, 113.

205. Poinar, G. O., Jr. and Thomas, G. M., Significance of *Achromobacter nematophilus* Poinar and Thomas (Achromobacteracae: Eubacteriales) in the development of the nematode, DD-136 (*Neoaplectana* sp. Steinernematidae), *Parasitology*, 56, 385, 1966.

206. Hoy, J. M., The use of bacteria and nematodes to control insects, *N.Z. Sci. Rev.*, 13, 56, 1955.

207. **Dumbleton, L. J.**, Bacterial and nematode parasites of soil insects, *N.Z. J. Sci. Technol. Sect. B*, 27, 76, 1945.

208. **Glaser, R. W.**, Continued culture of a nematode parasitic in the Japanese beetle, *J. Exp. Zool.*, 84, 1, 1940.

209. **McCoy, E. E. and Glaser, R. W.**, Nematode culture for Japanese beetle control, *N.J. Agric.*, 265, 3, 1936.

210. **McCoy, E. E. and Girth, H. B.**, The culture of *Neoaplectana glaseri* on veal pulp, *N.J. Agric.*, 285, 3, 1938.

211. **Stoll, N. R.**, Conditions favoring the axenic culture of *Neoaplectana glaseri,* a nematode parasite of certain insect grubs, *Ann. N.Y. Acad. Sci.*, 77, 126, 1959.

212. **Glaser, R. W. and Farrell, C. C.**, Field experiments with the Japanese beetle and its nematode parasite, *J. N.Y. Entomol. Soc.*, 43, 345, 1935.

213. **Hoy, J. M.**, The use of bacteria and nematodes to control insects, *N.Z. Sci. Rev.*, 13, 56, 1955.

214. **Travassos, L.**, Una nova especie do genero *Neoaplectana Steiner,* 1929 (Nematoda), *Bol. Biol.*, 19, 150, 1931.

215. **Travassos, L.**, Una specie del genere *Neoaplectana* Steiner (Nematoda-Oxyuridae) parassita del *Conorrhynchus (Cleonus) mendicus* Gyll. (Coleoptera — Curculionidae), *Boll. Laborat. Zool. Gen. Agraria Inst. Sup. Agrar. Portici,* 26, 115, 1932.

216. **Stanuszek, S.**, Adaptation of studies on interspecies competition for differentiating nematode species of the genus *Neoaplectana,* Steiner, 1929 (Rhabditoidea; Steinernematidae) in Abstr. 11th Int. Symp. Nematology, 1972, 68.

217. **Stanuszek, S.**, *Neoaplectana feltiae* Filipjev, 1934 — a facultative parasite of the caterpillars of Agrotinae in Poland, *Zesz. Probl. Postepow Nauk Roln.*, 92, 355, 1970.

218. **Bovien, P.**, Some types of associations between nematodes and insects, *Vidensk. Medd. Dan Naturhist. Foren. Khobenhavn,* 101, 114, 1937.

219. **Galser, R. W., McCoy, E. E., and Girth, H. B.**, The biology and culture of *Neoaplectana chresima,* a new nematode parasitic in insects, *J. Parasitol.*, 28, 123, 1942.

220. **Hoy, J. M.**, The biology and host range of *Neoaplectana leucaniae,* a new species of insect-parasite nematode, *Parasitology,* 44, 392, 1954.

221. **Weiser, J., and Köhler, W.**, Hlistice (Nematoda) Jako cizopasnici larev ploskohrbetky, *Acantholyda nemoralis* Thoms. v Polsku, *Cesk. Parazitol.*, 2, 185, 1955.

222. **Weiser, J.**, *Neoaplectana carpocapsae* n. sp. (Anguillulata, Steinernematinae) nový cizopasník housenek obaleče jablečného, *Carpocapsa pomonella* L., *Vestn. Cesk. Spol. Zool.*, 19, 44, 1955.

223. **Dutky, S. R. and Hough, W. S.**, Note on a parasitic nematode from codling moth larvae, *Carpocapsae pomonella* (Lepidoptera, Olethreutidae), *Proc. Entomol. Soc. Wash.*, 57, 244, 1955.

224. **Poinar, G. O., Jr.**, The presence of *Achromobacter nematophilus* in the infective stage of a *Neoaplectana* sp. (Steinernematidae: Nematoda) *Nematologica,* 12, 105, 1966.

225. **Hansen, E. L., Yarwood, E. A., Jackson, G. J., and Poinar, G. O., Jr.**, Axenic culture of *Neoaplectana carpocapsae* in liquid media, *J. Parasitol.*, 54, 1236, 1968.

226. **Caltagirone, L.**, personal correspondence, 1963.

227. **All, J.**, personal correspondence, 1977.

228. **Veremchuk, G. V.**, New species of entomopathogenic nematodes of the genus *Neoaplectana* (Rhabditida: Steinernematida) from wireworms (Elateridae), in Mat. Sci. Conf. All Union Soc. Helminthologists, Part 1, Moscow, 1969, 44.

229. **Poinar, G. O., Jr. and Himsworth, P. T.**, *Neoaplectana* parasitism of larvae of the greater wax moth, *Galleria mellonella, J. Invertebr. Pathol.,* 9, 241, 1967.

230. **Poinar, G. O., Jr. and Leutenegger, R.**, Anatomy of the infective and normal third-stage juveniles of *Neoaplectana carpocapsae* Weiser (Steinernematidae: Nematoda), *J. Parasitol.*, 54, 340, 1968.

231. **Welch, H. E., and Bronskill, J. F.**, Parasitism of mosquito larvae by the nematode, DD-136 (Nematoda: Neoaplectanidae), *Can. J. Zool.*, 40, 1263, 1962.

232. **Triggiani, O. and Poinar, G. O., Jr.**, Infection of adult Lepidoptera by *Neoaplectana carpocapsae* (Nematoda), *J. Invertebr. Pathol.*, 27, 413, 1976.

233. **Poinar, G. O., Jr., van der Geest, L., Helle, W., and Wassink, H.**, Experiments with organisms from hosts other than *Glossina,* in *Tsetse,* Laird, M., Ed., International Development Research Center, Canadian Research Development Institute, Pub. No. IDRC-077E, Ottawa, 1977, 88.

234. **Schmiege, D. C.**, The feasibility of using a neoaplectanid nematode for control of some forest insect pests, *J. Econ. Entomol.*, 56, 427, 1963.

235. **Reed, E. M. and Wallace, H. R.**, Leaping locomotion by an insect-parasitic nematode, *Nature (London),* 206, 210, 1965.

236. **Kaya, H.**, Development of the DD-136 strain of *Neoaplectana carpocapsae* at constant temperatures, *J. Nematol.*, 9, 346, 1977.

237. **Dutky, S. R.**, Insect microbiology, *Adv. Appl. Microbiol.*, 1, 175, 1959.

238. Welch, H. E. and Briand, L. J., Field experiment on the use of a nematode for the control of vegetable crop insects, *Proc. Entomol. Soc. Ont.*, 91, 197, 1961.

239. Kamionek, M., Maslana, I., and Sandner, H., The survival of invasive larvae of *Neoaplectana carpocapsae* Weiser in a waterless environment under various conditions of temperature and humidity, *Zesz. Probl. Postepow Nauk Roln.*, 154, 409, 1974.

240. Moore, G. E., The bionomics of an insect-parasitic nematode, *J. Kans. Entomol. Soc.*, 38, 101, 1965.

241. Simons, W. R. and Poinar, G. O., Jr., The ability of *Neoaplectana carpocapsae* (Steinernematidae: Nematodea) to survive extended periods of desiccation, *J. Invertebr. Pathol.*, 22, 228, 1973.

242. Webster, J. M. and Bronskill, J. F., Use of Gelgard M and an evaporation retardant to facilitate control of larch sawfly by a nematode-bacterium complex, *J. Econ. Entomol.*, 61, 1370, 1968.

243. Bedding, R. A., New methods increase the feasibility of using *Neoaplectana* spp. (Nematoda) for the control of insect pests, in *Proc. 1st Int. Colloq. Invertebrate Pathology,* Kingston, Canada, 1976. 250.

244. Reed, E. M. and Carne, P. B., The suitability of a nematode (DD-136) for the control of some pasture insects, *J. Invertebr. Pathol.*, 9, 196, 1967.

244a. El-Sherif, unpublished data.

244b. Gaugler, and Boush, personal correspondence.

245. Poinar, G. O., Jr. and Thomas, G. M., A new bacterium, *Achromobacter nematophilus* sp. nov. (Achromobacteriaeceae: Eubacteriales) associated with a nematode, *Int. Bull. Bacteriol. Nomencl. Taxon.*, 15, 249, 1965.

246. Poinar, G. O., Jr., Thomas, G. M., Veremchuk, G. V., and Pinnock, D. E., Further characterization of *Achromobacter nematophilus* from American and Soviet populations of the nematode *Neoaplectana carpocapsae* Weiser, *Int. J. Syst. Bacteriol.*, 21, 78, 1971.

247. Poinar, G. O., Jr. and Thomas, G. M., The nature of *Achromobacter nematophilus* as an insect pathogen, *J. Invertebr. Pathol.*, 9, 510, 1967.

248. Dutky, S. R., Thompson, J. V., and Cantwell, G. E., A technique for the mass propagation of the DD-136 nematode, *J. Insect Pathol.*, 6, 417, 1964.

249. White, G. F., A method for obtaining infective nematode larvae from cultures, *Science,* 66, 302, 1927.

250. Carne, P. B. and Reed, E. M., A single apparatus for harvesting infective stage nematodes emerging from their insect hosts, *Parasitology,* 54, 551, 1964.

251. House, H. L., Welch, H. E., and Cleugh, T. R., Food medium of prepared dog biscuit for the mass-production of the nematode DD-136 (Nematoda: Steinernematidae), *Nature (London),* 206, 847, 1965.

252. Hansen, E. L. and Cryan, W. S., Continuous axenic culture of free-living nematodes, *Nematologica,* 12, 138, 1966.

253. Veremchuk, G. V., personal correspondence, 1971.

254. Singh, J. and Bardhan, A. K., Effectiveness of DD-136, an entomophilic nematode against insect pests of agricultural importance, *Curr. Sci.,* 43, 622, 1974.

255. Lotkin, Yu. G. and Ivanova, S. G., A parasitic nematode — *Neoaplectana* in the control of insect pests, *Tr. Sakhalin. Obl. Stn. Zashch. Rast.,* 7, 55, 1970.

256. Creighton, C. S., Cuthbert, F. R., Jr., and Reid, W. J., Jr., Susceptibility of certain coleopterous larvae to the DD-136 nematode, *J. Invertebr. Pathol.,* 10, 368, 1968.

257. Welch, H. E., Test of a nematode and its associated bacterium for control of the Colorado potato beetle, *Leptinotarsa decemlineata* (Say), *Rep. Entomol. Soc. Ontario,* 88, 53, 1958.

258. Dutky, S. R., Thompson, J. V., and Hough, W. S., A Promising Nematode and the Associated Pathogen for Controlling Insect Pests, Entomology Res. Branch Circ., Bethesda, Maryland, 1956, 4.

259. Jourdheuil, J., Laumond, C., Bonifassi, E., and Millot, P., Recherches preliminaires sur l'utilisation des nematodes en vue de la lutte contre les insectes des crucifères, *J. Intern. Colza.,* May, 364, 1970.

260. Tang, J. L., Notas generales sobre nematodes portadores de bacterias como un metodo de control biologico, *Rev. Peru. Entomol.,* 1, 19, 1958.

261. Tedders, W. L., Weaver, D. J., and Wehunt, E. J., Pecan weevil: suppression of larvae with the fungi *Metarrhizium anisopliae* and *Beauveria bassiana* and the nematode *Neoaplectana carpocapsae, J. Econ. Entomol.,* 66, 723, 1973.

262. Harlan, D. P., Dutky, S. R., Padgett, G. R., Mitchell, J. A., Shaw, Z. A., and Bartlett, F. J., Parasitism of *Neoaplectana dutkyi* in white-fringed beetle larvae, *J. Nematol.,* 3, 280, 1971.

263. Lindegren, J. E., Current Prospects for Microbial Control of Nitidulid Beetles, *Res. Proc. Calif. Fig Inst.,* Fresno, May, 15, 1976.

264. Balthasar, V., *Monographie der Scarabaeidae und Aphodiidae der Palaearktischen und Orientalischen Region,* Vol. 1, Czechoslovakia Academy of Sciences, Prague, 1963, 391.

265. **Amaya, L. M. and Bustamante, E.,** Control microbiologico de tres especies de coleópteros plagas del suelo en Colombia, *Revista (Colombia),* 10, 269, 1975.

266. **Moore, G. E.,** *Dendroctonus frontalis* infection by the DD-136 strain of *Neoaplectana carpocapsae* and its bacterium complex, *J. Nematol.,* 2, 341, 1970.

267. **Finney, J. R. and Mordue, W.,** The susceptibility of the elm bark beetle *Scolytus scolytus* to the DD-136 strain of *Neoaplectana* sp., *Ann. Appl. Biol.,* 83, 311, 1976.

268. **Sandner, H.,** The Role of Nematodes as Factors Reducing Populations of Insect Pests, *Inst. Ecology Rep.,* Warsaw, 1972, 47.

269. Res. Branch Rep. for 1971, Canada Department of Agriculture, Ottawa, 1972, 377.

270. **Webster, J. M.,** Manipulation of environment to facilitate use of nematodes in biocontrol of insects, *Exp. Parasitol.,* 33, 197, 1973.

271. **Skierska, B. and Szadziewska, M.,** Laboratory tests for usability of the entomophilic nematodes Steinernematidae Chitwood et Chitwood 1937 in biological control of some noxious arthropods, *Bull. Inst. Mar. Trop. Med. Gdansk,* 27, 207, 1976.

272. **Hackett, K. J. and Poinar, G. O., Jr.,** The ability of *Neoaplectana carpocapsae* Weiser (Steinernematidae: Rhabditoidea) to infect adult honeybees (*Apis mellifera,* Apidae: Hymenoptera,), *Am. Bee J.* 113, 100, 1973.

273. **Kermarrec, A.,** Etude des relations synecologiques entre les nematodes et la fourmi-manioc, *Acromyrmex octospinosus* Reich., *Ann. Zool.,* 7, 27, 1975.

274. **Poinar, G. O., Jr. and Ennik, F.,** The use of *Neoaplectana carpocapsae* (Steinernematidae: Rhabditoidea) against adult yellowjackets (*Vespula* spp. Vespidae: Hymenoptera), *J. Invertebr. Pathol.,* 19, 331, 1972.

275. **Reese, K. M.,** Navy fights Formosan termite in Hawaii, *Chem. Eng. News,* p. 52, October 1971.

276. **Curtis, B.,** Report on the 4th Annual Almond board research conference *Almond Facts,* 42, 13, 1977.

277. **Jaques, R. P.,** Mortality of five apple insects induced by the nematode DD-136, *J. Econ. Entomol.,* 60, 741, 1967.

278. **Laumond, C.,** Utilisation practique des Neoaplectanidae contre divers Lépitopteres des cultures maraîchères, in *Abstr. 11th Int. Symp. Nematology,* Reading, England, 1972, 41.

279. **Israel, P., Rao, Y. R., Rao, P. S., and Varma, A.,** Control of paddy cut worms by DD-136, a parasitic nematode, *Curr. Sci.,* 38, 390, 1969.

280. **Tanada, Y. and Reiner, C.,** The use of pathogens in the control of the corn earworm, *Heliothis zea* (Boddie), *J. Insect. Pathol.,* 4, 139, 1962.

281. **Torii, T.,** Feasibility of using the entomophilic nematode, DD-136 as a biotic insecticide against rice stem borers, in *Approaches to Biological Control,* Yasumatsu, K. and Mori, H., Eds., University of Tokyo Press, Tokyo, 1975, 87.

282. **Tanada, Y. and Reiner, C.,** Microbial control of the artichoke plume moth, *Platyptilia carduidactyla* (Riley) (Pterophoridae, Lepidoptera), *J. Insect. Pathol.,* 2, 230, 1960.

283. **Rao, V. P. and Manjunath, T. M.,** DD-136: nematode that can kill many insect pests, *Indian Farming,* 16, 43, 1966.

284. **Rao, Y. R., Rao, P. S., Verma, A., and Israel, P.,** Tests with an insect parasitic nematode DD-136 (Nematoda: Steinernematidae) against the rice stem borer, *Tryporyza incertulas* Walker, *Indian J. Entomol.,* 33, 215, 1971.

285. **Berlowitz, A.,** personal communication, 1978.

286. **Lindegren, J.,** personal communication, 1978.

287. **Chamberlin, F. S. and Dutky, S. R.,** Tests of pathogens for the control of tobacco insects, *J. Econ. Entomol.,* 51, 560, 1958.

288. **Welch, H. E. and Briand, L. J.,** Tests of the nematode DD-136 and an associated bacterium for control of the Colorado potato beetle, *Leptinotarsa decemlineata* (Say), *Can. Entomol.,* 93, 759, 1961.

289. **Fox, C. J. S. and Jaques, R. P.,** Preliminary observations on biological insecticides against imported cabbageworm, *Can. J. Plant Sci.,* 46, 497, 1966.

290. **Jaques, R. P., Stultz, H. T., and Huston, F.,** The mortality of the pale apple leafroller and winter moth by fungi and nematodes applied to the soil, *Can. Entomol.,* 100, 183, 1968.

291. **Nash, R. F. and Fox, R. C.,** Field control of the Nantucket pine tip moth by the nematode DD-136, *J. Econ. Entomol.,* 62, 660, 1969.

292. **Yadava, C. P. and Rao, Y. S.,** On the effectiveness of the entomophilic nematode DD-136 in the biological control of insect pests of rice, *Oryza J. Assoc. Rice Res. Work.,* 7, 131, 1970.

293. **Cheng, H. H. and Bucher, G. E.,** Field comparison of the neoaplectanid nematode DD-136 with diazinon for control of *Hylemya* spp. on tobacco, *J. Econ. Entomol.,* 65, 1761, 1972.

294. **Landszabal, A., Fernandez, J., and Figueroa, P. A.,** Control biologico de *Spodoptera frugiperda* (J. E. Smith), con el nematodo *Neoaplectana carpocapsae* en maiz *(Zea mays), Acta Agron. (Palmira),* 23, 41, 1973.

295. **Lam. A. B. Q. and Webster, J. M.**, Effect of the DD-136 nematode and of a B-exotoxin preparation of *Bacillus thuringiensis* var *thuringiensis* on leatherjackets, *Tipula paludosa* larvae, *J. Invertebr. Pathol.*, 20, 141, 1972.

296. **Kirjanova, E. S. and Putschkova, L. B.**, A new parasite of the beet weevil *Neoaplectana bothynoderi* Kirjanova and Putschkova sp. n. (Nematoda), *Tr. Zool. Inst. Akad. Nauk SSSR*, 18, 53, 1955.

297. **Weiser, J.**, Ein neuer Nematode als Parasit der Engerlinge des Maikäfers *Melolontha melolontha* in der Tschechoslowakei, in Trans. 1st Int. Conf. Insect. Pathology and Biological Control, Prague, 1958, 331.

298. **Kakulia, G. A. and Veremchuk, G. V.**, *Neoaplectana georgica*, n. sp. (Nematoda: Steinernematidae) in *Amphimallon solstitialis*, *Soobshch. Akad. Nauk. Gruz. SSR*, 40, 713, 1965.

299. **Kurashvili, B. E. and Kakulia, G. A.**, Cultivation of *Neoaplectana georgica* Kakulia and Veremchuk, 1968, in Proc. 3rd Int. Congr. Parasitology Tblisi, Georgia, U.S.S.R., 465, 1974.

300. **Artyukhovsky, A. K.**, *Neoaplectana arenaria* n. sp. (Steinernematidae: Nematodae) from the may beetle in the Voronezh area, *Tr. Voronezh. Gos. Zapov.*, 15, 94, 1967.

301. **Veremchuk, G. V.**, New species of *Neoaplectana* (Rhabditida: Steinernematidae), nematodes pathogenic to Elateridae, *Mater. Nauchn. Konf. Ova. Gel'mintol.*, 1, 44, 1969.

302. **Turco, C. P.**, *Neoaplectana hoptha*, sp. n. (Neoaplectanidae: Nematoda), A parasite of the Japanese beetle, *Popillia japonica* Newm., *Proc. Helminthol. Soc. Wash.*, 37, 119, 1970.

303. **Kahn, A., Brooks, W. M., and Hirschmann, H.**, *Chromonema heliothidis* n. gen., n. sp. (Steinernematidae, Nematoda), a parasite of *Heliothis zea* (Noctuidae, Lepidoptera) and other insects, *J. Nematol.*, 8, 159, 1976.

304. **Veremchuk, G. V. and Litvinchuk, L. N.**, Nematodes of the genus *Neoaplectana* (Rhabditida: Steinernematidae) parasites of the larch looper moth, *Semiothisa pumila* Kusn. (Lepidoptera, Geometridae), in *New and Little Known Species of the Siberian Fauna*, Nauka, Novosibersk, 1971, 92.

305. **Poinar, G. O., Jr.**, Description and biology of a new insect parasitic rhabditoid, *Heterorhabditis bacteriophora* n. gen., n. sp. (Rhabditida: Heterorhabditidae n. fam.), *Nematologica*, 21, 463, 1975.

306. **Pereira, C.**, *Rhabditis hambletoni* n. sp., nema apparentemente semiparasito da "broca do algodoeiro" *(Gasterocercodes brasiliensis)*, *Arch. Inst. Biol. (São Paulo)*, 8, 215, 1937.

307. **Hamm, J. J.**, personal correspondence, 1977.

308. **Littig, K. S. and Swain, R. B.**, Studies on nematode 41088, a nematode parasite of the white-fringed beetles, in Special Rep. White-Fringed Beetle Investigations, Gulfport, Miss., 1943, 17.

309. **Laumond, C.**, Alternance de generations sexuées et parthénogénétiques dans le cycles d'un rhabditide entomoparasite nouveau vecteur de bactéries, in Abstr. 12th Int. Symp. Nematology, Granada, 1974, 60.

310. **Bedding, R. A. and Akhurst, R. J.**, Nematode control of insects, *C.S.I.R. Annu. Rep. 1975— 1976*, p. 71, 1976.

311. **Wouts, W.**, personal communication, 1977.

312. **Rogers, C.**, personal correspondence, 1975.

313. **Milstead, J. E. and Poinar, G. O., Jr.**, A new entomogenous nematode for pest management systems, *Calif. Agric.*, 32, 12, 1978.

314. **van Bracht, W.**, unpublished data, 1975.

315. **Poinar, G. O., Jr., Thomas, G. M., and Hess, R.**, Characteristics of the specific bacterium associated with *Heterorhabditis bacteriophora* (Heterorhabditidae: Rhabditida), *Nematologica*, 23, 97, 1977.

316. **Kahn, A. and Brooks, W. M.**, A chromogenic bioluminescent bacterium associated with the entomophilic nematode *Chromonema heliothidis*, *J. Invertebr. Pathol.*, 29, 253, 1977.

317. **Kaya, H. K.**, Infectivity of *Neoaplectana carpocapsae* and *Heterorhabditis* to pharate pupae of the gregarious insect parasite, *Apanteles militaris* in press.

318. **Laumond, C.**, Hétérogonie et adaptations morphologiques chez un Sphaerulariidae (Nematoda), parasite de *Baris caerulescens*, *C. R. Acad. Sci.*, 271, 1575, 1970.

319. **Bedding, R. A.**, *Deladenus wilsoni* n. sp. and *D. siricidicola* n. sp. (Neotylenchidae), entomophagous-mycetophagous nematodes parasitic in siricid woodwasps, *Nematologica*, 14, 515, 1968.

320. **Bedding, R. A.**, Biology of *Deladenus siricidicola* (Neotylenchidae) an entomophagous-mycetophagous nematode parasitic in siricid woodwasps, *Nematologica*, 18, 482, 1972.

321. **Taylor, K. L.**, The introduction and establishment of insect parasitoids to control *Sirex noctilio* in Australia, *Entomophaga*, 21, 429, 1976.

322. **Bedding, R. A. and Akhurst, R. J.**, Parasitisation of siricid woodwasps and associated hymenopterous parasites by *Deladenus* species (Nematoda: Neotylenchidae). I. Geographical distribution and host preference, *Nematologica*, in press.

323. **Bedding, R. A. and Akhurst, R. J.**, Use of the nematode *Deladenus siricidicola* in the biological control of *Sirex noctilio* in Australia, *J. Aust. Entomol. Soc.*, 13, 129, 1974.

324. **Nickle, W. R.**, *Heterotylenchus autumnalis* sp. n. (Nematoda: Sphaerulariidae), a parasite of the face fly, *Musca autumnalis* De Geer, *J. Parasitol.*, 53, 398, 1967.

325. **Stoffolano, J. G., Jr. and Nickle, W. R.**, Nematode parasite (*Heterotylenchus* sp.) of face fly *(Musca autumnalis)* in New York State, *J. Econ. Entomol.*, 59, 221, 1966.

326. **Stoffolano, J. G., Jr.**, Distribution of the nematode, *Heterotylenchus autumnalis*, a parasite of the face fly, in New England with notes on its origin, *J. Econ. Entomol.*, 61, 861, 1968.

327. **Világiová, I.**, *Heterotylenchus autumnalis* Nickle (1967) — a parasite of pasture flies, *Biologia, (Bratislava)*, 23, 397, 1968.

328. **Stoffolano, J. G., Jr. and Streams, F. A.**, Host reactions of *Musca domestica, Orthellia caesarion* and *Ravinia l'herminieri* to the nematode *Heterotylenchus autumnalis*, *Parasitology*, 63, 195, 1971.

329. **Stoffolano, J. G., Jr.**, Maintenance of *Heterotylenchus autumnalis* a nematode parasite of the face fly, in the laboratory, *Ann. Entomol. Soc. Am.*, 66, 469, 1973.

330. **Anon.**, Face Fly-Distribution and Use as Biological Control (Beetle and Nematodes in California), Coop. Econ. Insect Rep. No. 19, U. S. Department of Agriculture, 1969, 723.

331. **Anon.**, Face Fly-Distribution and Use as Biological Control (Beetle and Nematodes in California), Coop. Econ. Insect Rep. No. 19, U. S. Department of Agriculture, 1969, 743.

332. **Nicholas, W. L. and Hughes, R. D.**, *Heterotylenchus* sp. (Nematoda: Sphaerulariidae), a parasite of the Australian bush fly, *Musca vetustissima*, *J. Parasitol.*, 56, 116, 1970.

333. **Riding, I. L. and Hague, N. G. M.**, Some observations on a tylenchiid nematode *Howardula* sp. parasitizing the mushroom phorid *Megasalia halterata* (Phoridae, Diptera), *Ann. Appl. Biol.*, 78, 205, 1974.

334. **Hussey, N. W.**, Biological control of mushroom pests — fact and fantasy, *Mushroom Growers Assoc. Bull.*, 238, 448, 1969.

335. **Remillet, M. and van Waerebeke, D.**, Description et cycle biologique de *Howardula madecassa* n. sp. et *Howardula truncati* n. sp. (Nematoda: Sphaerulariidae) parasites de *Carpophilus* (Coleoptera: Nitidulidae), *Nematologica*, 21, 192, 1975.

336. **Lindegren, J. E.**, The Biology of *Howardula* sp., a Nematode Parasite of *Carpophilus mutilatus* (Coleoptera: Nitidulidae), M.S. thesis, Fresno State College, Fresno, Calif., 1970.

337. **Cobb, N. A.**, *Howardula benigna:* a nema parasite of the cucumber beetle, *Science*, 54, 667, 1921.

338. **Poinar, G. O., Jr. and van der Laan, P. A.**, Morphology and life history of *Sphaerularia bombi*, *Nematologica*, 18, 239, 1972.

339. **Bovien, P.**, On a new nematode, *Scatonema wülkeri* gen. et sp. n. parasitic in the body cavity of *Scatopse fuscipes* Meig. (Diptera nematocera), *Vidensk. Medd. Dan. Naturhist. Foren. Khobenhavn*, 94, 13, 1932.

340. **Bovien, P.**, *Proatractonema sciarae* n. g., n. sp., a parasitic nematode from the body cavity of a dipterous larvae, *Vidensk. Medd. Dan. Naturhist. Foren. Khobenhavn*, 108, 1, 1944.

341. **Poinar, G. O., Jr.**, The bionomics and parasitic development of *Tripius sciarae* (Bovien) (Sphaerulariidae: Aphelenchoidea), a nematode parasite of sciarid flies (Sciaridae: Diptera), *Parasitology*, 55, 559, 1965.

342. **Leuckart, K. G. F. R.**, Neue Beiträge zur Kenntniss des Baues und der Lebensgeschichte der Nematoden, *Abh. Saechs. Akad. Wiss. Leipzig Math. Naturwiss, Kl.*, 13, 565, 1887.

343. **Poinar, G. O., Jr. and Doncaster, C. C.**, The penetration of *Tripius sciarae* (Sphaerulariidae: Aphelenchoidea) into its insect host, *Bradysia paupera* Tuom. (Mycetophilidae: Diptera), *Nematologica*, 11, 73, 1965.

344. **Wachek, F.**, Die entoparasitischen Tylenchiden, *Parasitol. Schriftenr.*, 3, 119, 1955.

345. **Parvez, Z.**, A Serological Study of the Bacterium *Achromobacter nematophilus* from Various Strains of the Nematode, *Neoaplectana carpocapsae* Weiser, Ph.D. thesis, University of California, Berkeley, 1974, 148.

346. **Poinar, G. O., Jr.**, Arthropod immunity to worms, in *Immunity to Parasitic Animals,* Vol. 1, Jackson, G., Herman, R., and Singer, I., Eds., Appleton-Century-Crofts, New York, 1969, 173.

347. **Poinar, G. O., Jr.**, Insect immunity to parasitic nematodes, in *Contemporary Topics in Immunobiology,* Cooper, E. L., Ed., Plenum Press, New York, 1976, 167.

348. **Poinar, G. O., Jr. and Leutenegger, R.**, Ultrastructural investigation of the melanization process in *Culex pipiens* (Culicidae) in response to a nematode, *J. Ultrastruct. Res.*, 36, 149, 1971.

349. **Götz, P.**, Die Einkapselung von Parasiten in der Hämolymphe von *Chironomus-Larven* (Diptera), *Zool. Anz. Suppl.*, 33, 610, 1969.

350. **Esser, R. P. and Sobers, E. K.**, Natural enemies of nematodes, *Proc. Soil Crop Sci. Soc. Fla.*, 24, 326, 1964.

351. **Stirling, G. R. and Mankau, R.**, Biological control of nematode parasites of citrus by natural enemies, in Proc. 1977 Int. Soc. Citriculture, Orlando, Florida, in press.

352. **Platzer, E. and MacKenzie-Graham, L. L.**, Predators of *Romanomermis culicivorax*, in Proc. Calif. Mosq. Vector Cont. Assoc., Riverside, California, 1978, 46.

353. **Steiner, G.**, Mermithids parasitic in the tea bug (*Helopeltis antonii* Sign.), *Meded. Proefstn. Thee, Buitenzorg*, 94, 3, 1925.

354. **Veremchuk, G. V. and Issi, I. V.**, The development of microsporidians of insects in the entomopathogenic nematode *Neoaplectana agriotos* (Nematodes, Steinernematidae), *Parazitologiya,* 4, 3, 1970.

355. **Poinar, G. O., Jr.**, Description and observations on a cuticular infection of *Thelastoma pterygoton* n. sp. (Thelostomatidae: Nematodea) from *Oryctes* spp. (Scarabaeidae: Coleoptera), *Proc. Helminthol. Soc. Wash.,* 40, 37, 1973.

356. **Poinar, G. O., Jr. and Hess, R.**, Virus-like particles in the nematode *Romanomermis culicivorax* (Mermithidae), *Nature (London),* 266, 256, 1977.

357. **Nolan, R. A.**, Physiological studies with the fungus *Saprolegnia megasperma* isolated from the freshwater nematode *Neomesomermis flumenalis, Can. J. Bot.,* 53, 3032, 1975.

357a. **Stirling, G. R. and Platzer, E.**, personal correspondence.

358. **Poinar, G. O., Jr., Leutenegger, R., and Thomas, G. M.**, On the occurrence of protein platelets in the pseudocoelom of a nematode (*Hydromermis* sp.; Mermithidae), *Nematologica,* 16, 348, 1970.

359. **Ignoffo, C. M., Petersen, J. J., Chapman, H. C., and Novotny, J. F.**, Lack of susceptibility of mice and rats to the mosquito nematode *Reesimermis nielseni* Tsai and Grundman, *Mosq. News,* 34, 425, 1974.

359a. **Nutrilite Products, Inc.**, personal correspondence, 1970.

359b. **Gaugler,** personal correspondence.

360. **Swenson, K. G.**, Infection of the garden symphylan, *Scutigerella immaculata,* with the DD-136 nematode, *J. Invertebr. Pathol.,* 8, 133, 1966.

361. **Leidy, J.**, Descriptions of three Filariae, *Proc. Acad. Nat. Sci. Philadelphia,* 5, 117, 1850.

361a. **Fuller,** personal correspondence, 1973.

362. **Stiles, C. W.**, The zoological characters of the roundworm genus *Filaria* Mueller, 1787, with a list of the thread worms reported for man, *Bull. Hyg.,* 34, 31, 1907.

363. **Leidy, J.**, Notes on some parasitic worms, *Proc. Acad. Nat. Sci. Philadelphia,* 27, 14, 1875.

364. **Leidy, J.**, On a filaria reported to have come from a man, *Proc. Acad. Nat. Sci. Philadelphia,* 27, 130, 1880.

365. **Stiles, C. W.**, A reexamination of the type specimen of *Filaria restiformis* Leidy, 1880 = *Agamomermis restiformis, Bull. Hyg.,* 40, 19, 1908.

365a. **Vacura,** personal correspondence, 1973.

366. **Levine, N. D.**, *Nematode Parasites of Domestic Animals and of Man,* Burgess Publishing, Minneapolis, 1968, 600.

367. **Faust, E. C.**, *Human Helminthology,* Lea and Febiger, Philadelphia, 1929, 616.

368. **Baylis, H. A.**, Notes on two gordiids and a mermithid said to have been parasitic in man, *Trans. R. Soc. Trop. Med. Hyg.,* 21, 203, 1927.

369. **Neveu-Lemaire, M.**, *Traité d'helminthologie médicale et vétérinaire,* Vigot Freres, Paris, 1936, 1515.

370. **Leon, L. A.**, Cuarto caso de infeccion humana por *Agamomermis,* Rev. Med. Trop. Parasitol. Bacteriol. Clin. Lab. (Habana), 12, 25, 1946.

371. **Chabaud, A. G. and Lanz, P.**, Pseudo-parasitisme de l'homme par *Agamomermis* sp., *Ann. Parasitol. Hum. Comp.,* 26, 376, 1951.

371a. **Prudhoe,** personal correspondence, 1975.

372. **Niklas, O. F.**, Die Nematoden DD-136 (*Neoaplectana* sp.) und *Neoaplectana carpocapsae* Weiser, 1955 (Rhabditoidea) als Insektenparasiten. Eine Literaturübersicht, *Mitt. Biol. Bundensanst. Land Forstwirtsch. Berlin Dahlem,* 124, 40, 1967.

373. **Niklas, O. F.**, Ergänzungen zum Literaturbericht über die Nematode DD-136 (*Neoaplectana carpocapsae* Weiser), "strain DD-136"; Rhabditida, *Nachrichtenbl. Dtsch. Pflanzenschutzdienst (Berlin),* 5, 71, 1969.

374. **Milum, V. G.**, A larval mermithid, *Mermis subnigrescens* Cobb, as a parasite of the honeybee, *J. Econ. Entomol.,* 31, 460, 1938.

375. **Chow, F. H.-C.**, Colonization of *Neoaplectana dutkyi* Jackson, and Its Effects on Fall Armyworm, *Spodoptera frugiperda* (Smith), M.S. thesis, University of Florida, Gainesville, 1972, 28.

376. **Veremchuk, G. V.**, Some results on the propagation of the nematode, *Neoaplectana* sp. on artificial media, in *Helminths of Man, Animals, Plants and Methods of Controlling Them,* Akad. Nauk., Moscow, 1963, 198.

377. **Jones, C. M. and Perdue, J. M.**, *Heterotylenchus autumnalis,* a parasite of the face fly, *J. Econ. Entomol.,* 60, 1393, 1967.

378. **Richardson, P. N., Hesling, J. J., and Riding, I. L.**, Life cycle and description of *Howardula husseyi* n. sp. (Tylenchida: Allantonematidae), a nematode parasite of the mushroom phorid *Megaselia halterata* (Diptera: Phoridae), *Nematologica,* 23, 217, 1977.

379. **Danilov, L. G.**, Susceptibility of wireworms to infection of the nematode *Neoaplectana carpocapsae* Weiser, 1955, race, *Agriotos; Bioll. Vsesoyunaya Nauchno Issled. Inst. Zash. Rost.,* 30, 54, 1976.

380. **Laumond, C. Mauleon, H., and Kermarrec, A.**, Donnees nouvelles sur le spectre d'hotes et le parasitisme du nematode entomophage *Neoaplectana carpocapsae, Entomophaga,* 24, 13, 1979.

381. **Ahmad, R.,** Studies on *Graphognathus lecoloma* (Boh.) (Col:Curculionidae) and its natural enemies in the central provinces of Argentina, *Tech. Bull. Comm. Inst. Biol. Control,* 17, 19, 1976.
382. **Ahmad, R.,** Investigations on the white-fringed weevils *Naupactus durius* (Boh.) and *Pantomorus auripes* Hustache (Col:Curculionidae) and their natural enemies in Argentina, *Tech. Bull. Comm. Inst. Biol.* Control, 17, 37, 1976.

INDEX

A

Abate®
  effect on parasites, 22—23
Abathymermis
  species
    life cycle, 13
Acalymma
  trivattatum
    parasites of, 19
  vittatum
    parasites of, 194
Acanthoceilidae
  classification, 5
Acantholyda
  nemoralis
    infection, 129—130
Acanthoscelides
  obtectus
    infection, 150
Acari
  resistance, 224
Acarina
  parasites, 188
Acheta
  assimilis
    infection, 154
  domestica
    infection, 154
Achroia
  grisella
    infection, 153
Achromobacter
  nematophilus
    agglutination titer, 204
    as bacterial associate, 103, 110, 124,
      142—143, 173
    biological characteristics, 200
    cultural characteristics, 200
    culture, 147
    fluorescent antibody staining, 205
    growth studies, 202
    morphology, 203
    mutualistic associations, 199, 201
  species
    mutualistic associations, 200
Acrididae
  infection, 70
Acridoidea
  infection, 71, 77
Acrolepia
  assectella
    field trials, 157
    infection, 152, 153
Acromyrmex

octospinosus
  infection, 152
Actinolaimidae
  classification, 4
Actinolaimoidea
  classification, 4
Actinomyxidia
  species
    pathogenicity, 216
Acuariidae
  classification, 6
Acuarioidea
  classification, 6
Adoretus
  species
    field tests, 117
Adenophorea
  classification, 3
Aedes
  aegypti
    as host, 47
    development of parasite, 25
    field trials, 42
    infection, 41, 55, 151
  atlanticus
    development parasite, 25
    ecology, 20
    rate of parasitism, 19
  calceatus
    infection, 41
  canadensis
    development of parasite, 25
  communis
    as host, 44—45, 46
  crucians
    ecology, 20
  dorsalis
    development of parasite, 25
    factors affecting parasitism, 20, 22
  fulgens
    infection, 41
  fulvus pallens
    development of parasite, 25
  haworthi
    infection, 41
  marshalli
    infection, 41
  metallicus
    infection, 41
  mitchellae
    development of parasite, 25
    rate of parasitism, 19
  nearcticus
    as host, 46
  nigripes
    as host, 46

274 Nematodes for Biological Control of Insects